NANOSTRUCTURED CARBON FOR ENERGY GENERATION, STORAGE, AND CONVERSION

AAP Research Notes on Nanoscience and Nanotechnology

NANOSTRUCTURED CARBON FOR ENERGY GENERATION, STORAGE, AND CONVERSION

Edited by

V. I. Kodolov, DSc
Omari Mukbaniani, DSc
Ann Rose Abraham, PhD
A. K. Haghi, PhD

AAP APPLE ACADEMIC PRESS

First edition published 2023

Apple Academic Press Inc.
1265 Goldenrod Circle, NE,
Palm Bay, FL 32905 USA

760 Laurentian Drive, Unit 19,
Burlington, ON L7N 0A4, CANADA

CRC Press
6000 Broken Sound Parkway NW,
Suite 300, Boca Raton, FL 33487-2742 USA

4 Park Square, Milton Park,
Abingdon, Oxon, OX14 4RN UK

Library and Archives Canada Cataloguing in Publication

Title: Nanostructured carbon for energy generation, storage, and conversion / edited by V.I. Kodolov, DSc, Omari Mukbaniani, DSc, Ann Rose Abraham, PhD, A.K. Haghi, PhD.
Names: Kodolov, Vladimir I. (Vladimir Ivanovich), editor. | Mukbaniani, O. V. (Omar V.), editor. | Abraham, Ann Rose, editor. | Haghi, A. K., editor.
Series: AAP research notes on nanoscience & nanotechnology.
Description: First edition. | Series statement: AAP research notes on nanoscience & nanotechnology | Includes bibliographical references and index.
Identifiers: Canadiana (print) 20220479062 | Canadiana (ebook) 20220479119 | ISBN 9781774911488 (hardcover) | ISBN 9781774911495 (softcover) | ISBN 9781003314967 (ebook)
Subjects: LCSH: Carbon nanotubes. | LCSH: Graphene. | LCSH: Nanostructured materials. | LCSH: Energy harvesting. | LCSH: Energy storage. | LCSH: Energy conversion. | LCSH: Renewable energy sources.
Classification: LCC TK2945.C37 N36 2023 | DDC 621.31/2—dc23

Library of Congress Cataloging-in-Publication Data

CIP data on file with US Library of Congress

ISBN: 978-1-77491-148-8 (hbk)
ISBN: 978-1-77491-149-5 (pbk)
ISBN: 978-1-00331-496-7 (ebk)

ABOUT THE AAP RESEARCH NOTES ON NANOSCIENCE & NANOTECHNOLOGY BOOK SERIES:

AAP Research Notes on Nanoscience & Nanotechnology reports on research development in the field of nanoscience and nanotechnology for academic institutes and industrial sectors interested in advanced research.

BOOKS IN THE AAP RESEARCH NOTES ON NANOSCIENCE & NANOTECHNOLOGY BOOK SERIES

- **Nanostructure, Nanosystems and Nanostructured Materials: Theory, Production, and Development**
  Editors: P. M. Sivakumar, PhD, Vladimir I. Kodolov, DSc, Gennady E. Zaikov, DSc, A. K. Haghi, PhD
- **Nanostructures, Nanomaterials, and Nanotechnologies to Nanoindustry**
  Editors: Vladimir I. Kodolov, DSc, Gennady E. Zaikov, DSc, and A. K. Haghi, PhD
- **Foundations of Nanotechnology:**
  Volume 1: Pore Size in Carbon-Based Nano-Adsorbents
  A. K. Haghi, PhD, Sabu Thomas, PhD, and Moein MehdiPour MirMahaleh
- **Foundations of Nanotechnology: Volume 2: Nanoelements Formation and Interaction**
  Sabu Thomas, PhD, Saeedeh Rafiei, Shima Maghsoodlou, and Arezo Afzali
- **Foundations of Nanotechnology: Volume 3: Mechanics of Carbon Nanotubes**
  Saeedeh Rafiei
- **Engineered Carbon Nanotubes and Nanofibrous Material: Integrating Theory and Technique**
  Editors: A. K. Haghi, PhD, Praveen K. M., and Sabu Thomas, PhD
- **Carbon Nanotubes and Nanoparticles: Current and Potential Applications**
  Editors: Alexander V. Vakhrushev, DSc, V. I. Kodolov, DSc, A. K. Haghi, PhD, and Suresh C. Ameta, PhD
- **Advances in Nanotechnology and the Environmental Sciences: Applications, Innovations, and Visions for the Future**
  Editors: Alexander V. Vakhrushev, DSc, Suresh C. Ameta, PhD, Heru Susanto, PhD, and A. K. Haghi, PhD
- **Chemical Nanoscience and Nanotechnology: New Materials and Modern Techniques**
  Editors: Francisco Torrens, PhD, A. K. Haghi, PhD, and Tanmoy Chakraborty, PhD

- **Nanomechanics and Micromechanics: Generalized Models and Nonclassical Engineering Approaches**
  Editors: Satya Bir Singh, PhD, Alexander V. Vakhrushev, DSc, and A. K. Haghi, PhD
- **Carbon Nanotubes: Functionalization and Potential Applications**
  Editors: Ann Rose Abraham, Soney C. George, PhD, and A. K. Haghi, PhD
- **Carbon Nanotubes for a Green Environment: Balancing the Risks and Rewards**
  Editors: Shrikaant Kulkarni, PhD, Iuliana Stoica, PhD, and A. K. Haghi, PhD
- **Nanostructured Carbon for Energy Generation, Storage, and Conversion**
  Editors: V. I. Kodolov, DSc, Omari Mukbaniani, DSc, Ann Rose Abraham, PhD, and A. K. Haghi, PhD

ABOUT THE EDITORS

V. I. Kodolov, DSc

Vladimir I. Kodolov, DSc, is professor and head of the Department of Chemistry and Chemical Technology at M. T. Kalashnikov Izhevsk State Technical University in Izhevsk, Russia, as well as Chief of Basic Research at the High Educational Center of Chemical Physics and Mesoscopy at the Udmurt Scientific Center, Ural Division, at the Russian Academy of Sciences. He is also the Scientific Head of the Innovation Center at the Izhevsk Electromechanical Plant in Izhevsk, Russia. He is vice editor-in-chief of the Russian journal *Chemical Physics and Mesoscopy* and is a member of the editorial boards of several Russian journals. He is the Honorable Professor of the M. T. Kalashnikov Izhevsk State Technical University, Honored Scientist of the Udmurt Republic, Honored Scientific Worker of the Russian Federation, Honorary Worker of Russian Education, and also Honorable Academician of the International Academic Society.

Omari Mukbaniani, DSc

Omari Mukbaniani, DSc, is a full professor at Ivane Javakhishvili Tbilisi State University (TSU), Faculty of Exact and Natural Sciences, Department of Chemistry; Chair of Macromolecular Chemistry, Tbilisi, Georgia. He is also Director of the Institute of Macromolecular Chemistry and Polymeric Materials at TSU and a member of the Academy of Natural Sciences of Georgia. For several years he was a member of the advisory board of the journal *Proceedings of Iv. Javakhishvili Tbilisi State University* (Chemical Series) and a contributing editor of the journal *Polymer News, Polymers Research Journal,* and *Chemistry and Chemical Technology.* His research interests include polymer chemistry, polymeric materials, and chemistry of organosilicon compounds. He is the author of more than 480 publications, 25 books, monographs, and 10 inventions. In 2007, he started the International Symposium on Polymers and Advanced Materials (ICSP&AM), which takes place every two years in Georgia. In 2018, he was a Chair of the 26th World Annual Forum on Advanced Materials PolyChar 26.

Ann Rose Abraham, PhD

Ann Rose Abraham, PhD, is currently an Assistant Professor at the Department of Physics, Sacred Heart College (Autonomous), Thevara, Kochi, Kerala, India. She has expertise in the field of condensed matter physics, nanomagnetism, multiferroics and polymeric nanocomposites, etc. She has research experience at various reputed national institutes, including Bose Institute, Kolkata, India; SAHA Institute of Nuclear Physics, Kolkata, India; UGC-DAE CSR Centre, Kolkata, India; and she has collaborated with various international laboratories. She is a recipient of a Young Researcher Award in physics and Best Paper Awards–2020, 2021. She served as Assistant Professor and Examiner at the Department of Basic Sciences, Amal Jyothi College of Engineering, under APJ Abdul Kalam Technological University, Kerala, India. Dr. Abraham is a frequent speaker at national and international conferences. She has a good number of publications to her credit in many peer-reviewed high impact journals of international repute. She has authored many book chapters and edited more than 20 books with Taylor and Francis, Elsevier, etc.

A. K. Haghi, PhD

A. K. Haghi, PhD, is the author and editor of over 250 books, as well as over 1000 published papers in various journals and conference proceedings. Dr. Haghi has received several grants, consulted for a number of major corporations, and is a frequent speaker to national and international audiences. Since 1983, he served as professor at several universities. He is the former Editor-in-Chief of the *International Journal of Chemoinformatics and Chemical Engineering* and *Polymers Research Journal* and is on the editorial boards of many international journals. He is also a member of the Canadian Research and Development Center of Sciences and Cultures. He holds a BSc in urban and environmental engineering from the University of North Carolina (USA), an MSc in mechanical engineering from North Carolina A&T State University (USA), a DEA in applied mechanics, acoustics and materials from the Université de Technologie de Compiègne (France), and a PhD in engineering sciences from Université de Franche-Comté (France).

CONTENTS

Contributors .... *xiii*

Abbreviations .... *xvii*

Preface .... *xxi*

1. **Carbon Nanotube/Graphene Hybrids for Energy Storage Applications 1**
Arunima Reghunadhan, K. C. Nimitha, and Jiji Abraham

2. **Graphene-based Materials for Energy Storage and Conversion Application 35**
Ashutosh Majhi, Lipsa Shubhadarshinee, Patitapaban Mohanty, Bigyan Ranjan Jali, Priyaranjan Mohapatra, and Aruna Kumar Barick

3. **Carbon Nanotube-Ferrite Hybrid Frameworks for Electromagnetic Wave Attenuation and other Potential Applications 87**
Ann V Sony, Amala Rose Augustine, Snitha Vinod K, and Ann Rose Abraham

4. **Heteroatom-Doped Graphene for Energy Storage Applications 119**
Jaison M. Joy, Mohammad Reza Saeb, Jyotishkumar Parameswaranpillai, and C. D. Midhun Dominic

5. **Carbon Nanotubes for Sustainable and Clean Energy Applications 145**
N. G. Divya, V. N. Anjana, and V. N. Archana

6. **Carbon-based Nanomaterials for Energy Generation and Storage Applications 175**
Arun Kumar K.V., Greeshma Sara John, Athira Maria Johnson, Arjun Suresh P., and Unnikrishnan N.V.

7. **Green Energy Applications of Graphene 213**
Vidya L., Aparna Raj, Neelima S., Riju K. Thomas, and C. Sudarsanakumar

Index* *239

CONTRIBUTORS

Ann Rose Abraham
Department of Physics, Sacred Heart College (Autonomous), Thevara, Kochi 682013, Kerala, India;
E-mail: annroseabraham@shcollege.ac.in

Jiji Abraham
Department of Chemistry, Vimala College (Autonomous) Thrissur 680009, Kerala, India

V. N. Anjana
Sree Sankara Vidyapeetom College, Valayanchirangara, Perumbavoor 683556, Kerala, India

V. N. Archana
Mar Athanasius College, Kothamangalam 686666, Kerala, India; E-mail: archanavn90@gmail.com

Amala Rose Augustine
Department of Physics, Sacred Heart College (Autonomous), Thevara, Kochi 682013, Kerala, India

Aruna Kumar Barick
Department of Chemistry, Veer Surendra Sai University of Technology, Siddhi Vihar, Burla,
Sambalpur 768018, Odisha, India; E-mail: akbarick_chem@vssut.ac.in

N. G. Divya
Cochin University of Science and Technology, Cochin 682022, Kerala, India

C. D. Midhun Dominic
Department of Chemistry, Sacred Heart College (Autonomous), Kochi 682013, Kerala, India;
E-mail: midhundominic@shcollege.ac.in

Bigyan Ranjan Jali
Department of Chemistry, Veer Surendra Sai University of Technology, Siddhi Vihar, Burla,
Sambalpur 768018, Odisha, India

Greeshma Sara John
Department of Physics, CMS College (Autonomous), Kottayam 686001, Kerala, India
Nanotechnology and Advanced Materials Research Centre, CMS College (Autonomous),
Kottayam 686001, Kerala, India

Athira Maria Johnson
Department of Physics, CMS College (Autonomous), Kottayam 686001, Kerala, India
Nanotechnology and Advanced Materials Research Centre, CMS College (Autonomous),
Kottayam 686001, Kerala, India

Jaison M. Joy
Department of Chemistry, Sacred Heart College (Autonomous), Kochi 682013, Kerala, India

Snitha Vinod K.
Department of Physics, Sacred Heart College (Autonomous), Thevara, Kochi 682013, Kerala, India

Vidya L.
School of Pure & Applied Physics, Mahatma Gandhi University, Kottayam, Kerala, India

Ashutosh Majhi
Department of Chemistry, Veer Surendra Sai University of Technology, Siddhi Vihar, Burla, Sambalpur 768018, Odisha, India

Patitapaban Mohanty
Department of Chemistry, Veer Surendra Sai University of Technology, Siddhi Vihar, Burla, Sambalpur 768018, Odisha, India

Priyaranjan Mohapatra
Department of Chemistry, Veer Surendra Sai University of Technology, Siddhi Vihar, Burla, Sambalpur 768018, Odisha, India

K. C. Nimitha
Department of Chemistry, Vimala College (Autonomous) Thrissur 680009, Kerala, India

Arjun Suresh P.
Department of Physics, CMS College (Autonomous), Kottayam 686001, Kerala, India
Nanotechnology and Advanced Materials Research Centre, CMS College (Autonomous), Kottayam 686001, Kerala, India

Jyotishkumar Parameswaranpillai
School of Biosciences, Mar Athanasios College for Advanced Studies Tiruvalla (MACFAST), Pathanamthitta 689101, Kerala, India

Aparna Raj
School of Pure & Applied Physics, Mahatma Gandhi University, Kottayam, Kerala, India

Arunima Reghunadhan
Department of Chemistry, TKM College of Engineering, Karicode, Kollam-5, Kerala, India;
E-mail: arunimarenjith02@gmail.com

Neelima S.
School of Pure & Applied Physics, Mahatma Gandhi University, Kottayam, Kerala, India

Mohammad Reza Saeb
Department of Polymer Technology, Faculty of Chemistry, Gdańsk University of Technology, Gabriela Narutowicza 11/12, 80-233 Gdańsk, Poland

Lipsa Shubhadarshinee
Department of Chemistry, Veer Surendra Sai University of Technology, Siddhi Vihar, Burla, Sambalpur 768018, Odisha, India

Ann V. Sony
Department of Physics, Sacred Heart College (Autonomous), Thevara, Kochi 682013, Kerala, India

C. Sudarsanakumar
School of Pure & Applied Physics, Mahatma Gandhi University, Kottayam, Kerala, India;
E-mail: c.sudarsan.mgu@gmail.com

Riju K. Thomas
Bharata Mata College, Thrikkakara, Ernakulam, Kerala, India

Unnikrishnan N.V.
School of Pure and Applied Physics, Mahatma Gandhi University, Kottayam, Kerala, India

Arun Kumar K. V.
Department of Physics, CMS College (Autonomous), Kottayam 686001, Kerala, India
Nanotechnology and Advanced Materials Research Centre, CMS College (Autonomous), Kottayam 686001, Kerala, India; E-mail: arunkumar@cmscollege.ac.in

ABBREVIATIONS

0D	zero dimension
1D	one-dimensional
2D	two-dimensional
3D	three-dimensional
AC	activated carbon
ACNT	amorphous carbon nanotube
AFC	alkaline fuel cell
AGH	aminated graphene honeycomb
AP	1-aminopyrene
BCP	bathocuproine
BET	Brunauer-Emmett-Teller
BF	barium ferrite
BHJ	bulk heterojunction
CB	conduction bands
CE	counter electrode
CEs	counter electrodes
CMG	chemically modified graphene
CNFs	carbon nanofibers
CNOs	carbon nano-onions
CNTs	carbon nanotubes
CuPc	copper phthalocyanine
CV	cyclic voltammetry
CVD	chemical vapor deposition
DAFCs	direct alcohol fuel cells
DFAFCs	direct formic acid fuel cells
DMC	dimethyl carbonate
DMFCs	direct methanol fuel cells
DOS	density of states
DSSCs	dye-sensitized solar cells
DWCNTs	double-walled CNTs
ECE	energy conversion efficiency
ECGnPs	edge-carboxylated graphene nanoplatelets
EDLC	electric double layer capacitor

EDX	energy-dispersive x-ray
EFGs	edge-functionalized graphene materials
EMI	electromagnetic interference
EPA	Environment Protection Agency
FCs	fuel cells
FED	field emission devices
FESEM	field emission scanning electron microscope
FETs	field-effect transistors
FFs	fill factors
f-RGO	functionalized RGO
FRET	resonance energy transfer
FTIR	Fourier transform infrared
FTO	F-doped SnO2
FTO	fluorine-doped tin oxide
GA	graphene aerogel
GC	glassy carbon
GNM	graphene nanomesh
GNP	graphene nanoplatelets
GNR	graphene nanoribbon
GNSs	graphene nanosheets
GO	graphene oxide
GPMN	graphene/PANI multilayered nanostructure
GPMN-W	GPMN in water
GQD	graphene quantum dot
GS	graphene sheets
HOM	higher-order mode
HOPG	high oriented pyrolytic graphite
HRP	horseradish peroxide
HRTEM	high-resolution transmission electron microscopy
HTL	hole-transport layer
IPCE	incident photons to current efficiency
ITO	Indium Tin Oxide
KIBs	potassium ion batteries
LBL	layer-by-layer
LIB	lithium-ion battery
Li-ion	lithium-ion
LLIP	liquid–liquid interfacial precipitation
LPE	liquid-phase exfoliation

LSBs	lithium–sulfur batteries
MB	methylene blue
MCFC	molten carbonate fuel cell
MFC	microbial fuel cell
MIT	Massachusetts Institute of Technology
MmNi3	Misch metal-based alloy hydride
Ms	magnetization
MWCNTs	multi-walled carbon nanotubes
MWS	microwave sintering
N-GFs	N-doped graphene foams
NGnPs	nitrogen edge-doped graphene nanoplatelets
NiCd	nickel–cadmium
Ni-MH	nickel metal hydride
NiO	oxide of nickel
NPs	nanoparticles
NZF	nickel zinc ferrite
OCV	open circuit voltage
OD	optical density
OPV	organic photovoltaic
ORR	oxygen reduction reaction
OSCs	organic solar cells
P3HT	poly3-hexylthiophene
P3OT	poly (3-octylthiophene)
PAHs	polycyclic aromatic hydrocarbons
PAFC	phosphoric acid fuel cell
PAni	polyaniline
PCE	power conversion efficiency
PDDA	polydiallyl dimethylammonium chloride
PDT	photodynamic therapy
PECVD	Plasma-enhanced CVD
PEI	polyethylene imine
PEMFC	proton exchange membrane fuel cells
PET	polyethylene terephthalate
PF	formaldehyde resin
PG	pristine graphene
PL	photoluminescence
PPy	polypyrrole
PSCs	polymer solar cells

PSS	polysodium 4-styrensulfonate
Pt	platinum
PTT	photothermal therapy
PV	photovoltaic
PVDF	poly(vinylidene fluoride)
QD	quantum dot
QDSCs	quantum dot solar cells
QHE	quantum hall effect
RAM	radar-absorbing materials
RBM	radial breathing mode
rGO	reduced graphene oxide
RhB	rhodamine B
RT	room temperature
SC	supercapacitor
SEM	scanning electron microscopy
Si	silicon
SIBs	sodium-ion batteries
SLS	sodium lignosulfonates
SOFC	solid-oxide fuel cell
SSA	specified surface area
SSCs	semi-transparent solar cells
STH	solar-to-hydrogen
SWCNT	single-wall carbon nanotube
TCO	transparent conducting oxide
TCs	transparent conductors
TCFs	transparent conductive film
TEM	transmission electron microscopy
TPC	transient photocurrent
TPV	transient photovoltage
VACNTs	vertically aligned carbon nanotubes
VAGNAs	vertically aligned graphene nanosheet arrays
VB	valence bands
VSM	vibrating sample magnetometer
XRD	X-ray diffractometry

PREFACE

This new volume presents the applications of different carbon nanomaterials and graphene-carbon-nanotubes hybrids for energy generation, energy storage, and energy conversion.

This book will be of interest to a wide readership in various fields of materials science and engineering and focuses solely on science and emerging applications of the rapidly advancing field of carbon nanomaterials with exceptional properties and characteristics.

The term "nanomaterials" is comprehended as the material with constituent particle size in the range of 1–100 nm in size. The field of materials synthesis, characterization, and applications are completely rejuvenated with this class of materials. These materials have been a central point of application research due to their novel properties resulting from their reduced dimensional variations as well their abilities to be building blocks for more complex nanostructures. To explain more clearly, here we have shown classification of nanomaterials in the broad categories on the basis of various parameters such as dimensionality, morphology, composition, uniformity, and agglomeration.

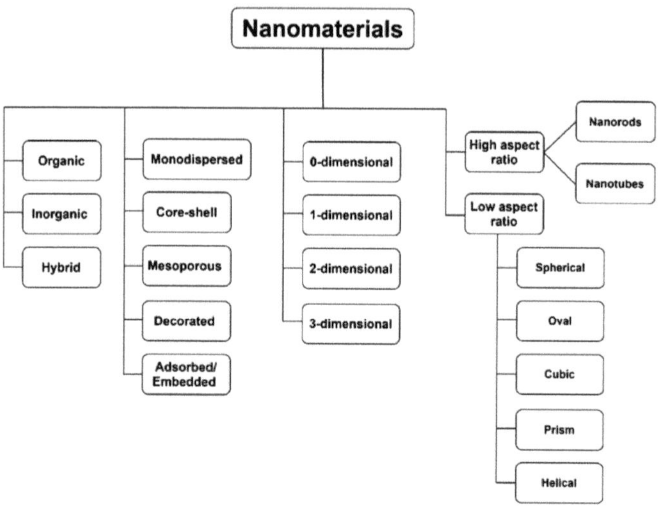

Nanomaterials and the science and technology of carbon nanotubes (CNTs) evolved on the basis of such materials are going to play a crucial role in the development of today's modern societies. The world of science and technology of carbon nanotubes are today in the avenues of scientific regeneration and scientific ingenuity.

However, there are many advantages associated with the use and application of nanomaterials.

Graphene and carbon nanotubes are both made of carbon atoms. Both graphene and carbon nanotubes have exceptional engineering properties, which can often be similar. These properties make carbon nanotubes attractive for engineering devices.

Carbon-based nanomaterials are produced and used in many industrial sectors. Carbon-based nanomaterials include carbon nanotubes (CNTs), nanographites, fullerenes, conducting carbon nanomaterials, nanodiamonds, graphene oxides, nanofibers, carbon black, carbon-onions, and hybrids like carbon nanotubes embedded into polymer composites.

This book presents a comprehensive overview of recent developments on carbon-based nanomaterials for energy generation and conversion, along with their storage applications.

CHAPTER 1

CARBON NANOTUBE/GRAPHENE HYBRIDS FOR ENERGY STORAGE APPLICATIONS

ARUNIMA REGHUNADHAN[1*,#], K. C. NIMITHA[2,#], and JIJI ABRAHAM[2,#]

[1]*Department of Chemistry, TKM College of Engineering, Karicode, Kollam-5, Kerala, India*

[2]*Department of Chemistry, Vimala College (Autonomous) Thrissur 680009, Kerala, India*

**Corresponding author. E-mail: arunimarenjith02@gmail.com*
#All the authors contributed equally to this chapter.

ABSTRACT

Energy demand has become a very serious issue in the 21st century. Nanomaterials with controlled aspect ratio and composition can provide options for developing electrical storage devices like batteries and capacitors. This chapter deals with carbon nanotube/graphene hybrids for energy storage applications. An overview of carbon nanotube/graphene hybrids, their characteristics, and synthetic strategies are incorporated in the first half. Later use of CNT–graphene hybrids in fuel cells as supercapacitors and as hybrids for solar cells was also reviewed in this chapter. The hybrid material shows a good synergistic effect which will give improved performance.

Nanostructured Carbon for Energy Generation, Storage, and Conversion. V. I. Kodolov, DSc, Omari Mukbaniani, DSc, Ann Rose Abraham, PhD, and A. K. Haghi, PhD (Eds.)

1.1 INTRODUCTION

Energy demand has become a very serious issue in the 21st century. Rapid consumption of fossil fuels faces a challenge in the current scenario. Fossil fuels also cause environmental pollution. As a result, there seems to be a need for renewable energy sources. These energy sources can effectively replace fossil fuels and contribute to the long-term sustainability of our ecosystem. This can promote the economic growth of our country too. Numerous studies are being conducted all over the world to create novel materials that can replace fossil fuels while also being environmentally friendly. The quest for energy storage devices has become very important since very ancient times. In ancient times, man uses wood and charcoal for generating fire. Coal which can be obtained from buried plants can store solar energy in a higher density than wood and charcoal. Coal was used in power engines and later in the production of electricity. Petroleum products that are generated from biodegraded organic materials were then used as an astounding storage medium for solar energy and are extensively used since the beginning of this millennium. The combustion of fossil fuels produces large amounts of toxic gases which can pollute the globe. The major drawback of using fossil fuels is the evolution of greenhouse gas like carbon dioxide which can alter the climate drastically.[1] Fossil fuels are non—renewable and their reserves are depleted now faster than ever before. These reasons highlight the demand for renewable and green energy sources and storage devices and this is found to be the major challenge that human beings are confronted with today.[2]

Electricity can be generated from fuel-burning power, wind energy, solar energy, hydropower, tide power, and nuclear power. With the drastic development of industries and increase in global population, the consumption of electrical energy found to get increased tremendously. Thus, the need for energy storage becomes more vital. In an ideal energy storage device, one form of energy is getting converted into another form in a storable form and is available when needed. Various studies are being carried out for the invention of materials with high energy density and power density. Engineering nanostructured materials with high potential enhances excellent storage capacity, high-rate capabilities, and long life spans.[3]

1.1.1 MATERIALS USED FOR ENERGY STORAGE

Electricity production from renewable sources offers fewer environmental hazards. The limitation of such sources is that they fluctuate independently from demand. Therefore, storage of these energy forms requires more attention. The main possible techniques for energy storage are mechanical, chemical, and thermal. Fuel cells are considered as best among the storage devices which can produce hydrogen through the electrolysis of water. Here, off-peak electricity uses to carry out electrolysis and produce hydrogen. A fuel cell uses this hydrogen and oxygen from the air to produce peak hour electricity. Fuel cells are of many types, such as alkaline fuel cells, polymer exchange membrane fuel cells, phosphoric acid fuel cells, molten carbonate fuel cells considering the electrolyte used, design, and field of application. Another important energy storage device is the supercapacitor. Supercapacitors have both the characteristics of capacitors and electrochemical batteries. They store energy in the form of an electric field created between two electrodes. Here, insulating materials in capacitors are replaced by an electrolyte ionic conductor. Ionic mobility takes place here along a conducting electrode. Flow batteries are other means of energy storage that consist of two electrodes and used chemical compounds in a liquid state for energy storage. Chemical storage can also be achieved by accumulators. These systems are having both storage and electricity release by alternating the charge–discharge phases.[4]

1.1.2 ROLE OF NANOMATERIALS IN ENERGY STORAGE

Electrical storage devices, such as batteries and capacitors can be fabricated with nanomaterials too. Redox-based supercapacitors with nanomaterial-based electrodes are showing high energy density and power capabilities. Challenges faced are the durability, cost of production, processability, and environmental effects. Major advantages of nanostructured electrodes in supercapacitors are the presence of larger electrode/electrolyte contact area due to the reduced dimensions, reduction in the ionic and electronic diffusion distance, enhanced tolerance to strain and structural distortion, and offering new properties, such as low weight, transparency, flexibility and biodegradability.[5]

Nanomaterials with controlled size, shape, and composition can provide options for developing highly active catalysts for fuel cell reactions. For example, Pt and Pt-based nanomaterials can act as perfect electrocatalysts in fuel cell applications. At the anode, it can successfully catalyze the oxidation of a molecule like a hydrogen, and at the cathode, it can catalyze the oxygen reduction reaction (ORR).[6] Major limitations of Pt-based electrocatalysts are the high price and poor utilization efficiency. Several studies have been carried out to replace Pt with a much more cheap and efficient catalyst. This can be achieved by the development of carbon-based nanomaterials, such as activated carbon, graphite nanofibers, carbon nanotubes (CNTs). In these materials, graphene has received much attention due to its unique electrical properties, surface area, and of course low cost. Graphene can effectively be used as an electrocatalyst, and it reduces Pt loading in fuel cells.[2] A single layer of graphene is a two-dimensional material consisting of one atom thick hexagonal carbon lattice with delocalized pi electrons. Properties of graphene are largely affected by the number of layers. Single-layer of graphene possesses quantum electronic properties. However, graphene oxide typically behaves as insulators as a result of breaking the long-range aromatic behavior of graphene.

Due to its outstanding electron transport capabilities and high carrier mobility, graphene has been used as an acceptor in solar devices. Graphene can easily disperse in organic solvent and can be used to prepare graphene/polymer bulk nanocomposites. Graphene has also been performed as the interface layer of heterojunction solar cells. Conventionally used materials face many challenges like high acidity and hygroscopicity which adversely affect the device performance. This problem can be resolved by using graphene which acts as a hole-transport layer (HTL). Graphene thin films can also act as electrode materials for solar cells. This is due to its high transparency, electrical conductivity, and flexibility. This can replace conventionally used Indium tin oxide (ITO)-based materials which face challenges like high cost, brittle nature and ion diffusion into polymer layers.[2]

CNTs are another class of carbon-based materials with almost a graphitic structure. The available two forms of CNTs are single-walled (SWCNTs) and multiwalled (MWCNTs) carbon nanotubes. SWCNTs are one atom thick and are formed by rolling a graphene sheet to form a tube. MWCNTs are having many layers of graphitic carbon. Both are metallic or semiconducting depending on the direction that the tubes are rolled. In both, electronic transport occurs ballistically. This property allows CNTs to carry high charge densities without energy dissipation.[7] Because of their large specific surface area and strong electrical and mechanical qualities, CNTs have been widely used as electrode materials for energy storage devices. Vertically aligned carbon nanotubes (VACNTs) are having well-defined structures and are highly pure. Hence, they have many applications in electronics. CNTs are widely used in supercapacitors and lithium-ion batteries. Covalent and non-covalent functionalization of graphene and CNTs can enhance their applications in energy storage.[8] Figure 1.1 represents the two different allotropes of carbon like (a) graphene and (b) single-walled carbon nanotube (SWCNT).[9]

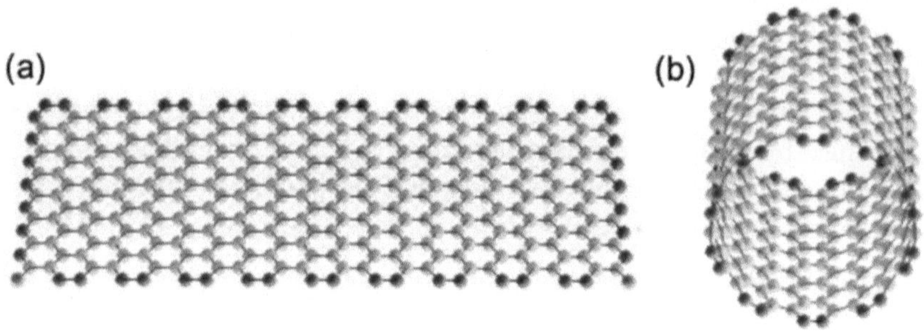

FIGURE 1.1 Carbon allotropes based on sp^2 carbon: (a) graphene, (b) single-walled carbon nanotube (Reproduced with permission from Ref. [9], Copyright 2011, Wiley).

Extended characteristics can be obtained from a covalently bonded graphene/SWCNT hybrid material. These hybrid materials can effectively be used in energy storage and nanoelectronic technologies. Hybrid materials exhibit a larger surface area. Covalent transformation of sp^2 carbon in between the planar graphene and SWCNT can be observed at the atomic resolution level. The enhanced properties of the hybrid materials pave way for the extraordinary applications of materials.[10]

1.2 GRAPHENE–CNT HYBRIDS

Graphene–CNT hybrids can be synthesized in many ways. Hybrids can form in various patterns too. Graphene-based hybrid nanostructures have been widely used for two applications mainly, that is, for transparent and flexible electrodes (TEs) and field-effect transistors (FETs).[11] One-dimensional CNTs can act as a spacer into the graphene oxide (GO) sheets and can behave like a three-dimensional carbon framework. It has the potential to improve its performance in dye-sensitized solar cells (DSSCs) as a counter electrode, in electrochemical capacitors, catalysts electrodes for methanol oxidation. It shows good adsorption of hydrogen due to extraordinary surface area.[12] Graphene–CNT aerogels can also be prepared which found astounding applications in several fields.

1.2.1 SYNTHESIS OF GRAPHENE—CNT HYBRIDS

Graphene—CNT nanotube hybrids can be synthesized through p-p interactions without any chemicals. The ratio between CNT and graphene has a great influence on the state and morphology of prepared materials. TEM study reveals that graphene nanosheets form a three-dimensional nanostructure by wrapping around individual CNTs. Prepared nanomaterials can be reduced by thermal treatment and are having a uniform structure. These materials possess good electrochemical performance in lithium-ion batteries. This can be applied as an anode in lithium-ion batteries and are having high Coulombic efficiency and extraordinary cyclability. There are several methods available for the synthesis of different types of graphene–CNT hybrids.[13]

One of the most common methods for the synthesis of CNTs is the catalytic chemical vapor deposition method (CVD). The reaction is carried out in a quartz tube under atmospheric pressure. The front part of the tube was first preheated and the main part is heated at a much more elevated temperature. This can be done by a horizontal furnace. A layer of pristine graphene was first homogeneously dispersed on the surface of a quartz plate. Carrier gases,such as argon, hydrogen are purged into it and the flow rates are controlled by electronic mass flow meters. Ferrocene dissolved in xylene can serve as a catalytic precursor. The solution is then carried into the reaction zone. Acetylene is provided to purify the system. After the process, the furnace was cooled down at room temperature. The resulting hybrids are further characterized by methods, such as Raman Spectroscopy, Scanning electron Spectroscopy, and transmission electron microscopy. It is found that the CNTs are well-aligned on the surface of graphene nanosheets. The thickness of graphene nanosheets are about few nanometers and the length of CNTs are in the range of 10 μm. Figure

1.2 shows the SEM images of graphene nanosheets–CNT hybrid materials prepared at two different temperatures, such as 550°C and 600°C. Studies show that low-growth temperatures can prevent damage of materials whereas high temperatures improve CNT yield. Therefore, the reaction is generally carried out under optimum temperature conditions.[14]

Agglomeration of the materials can be prevented by the ultrasonication method. In this method, the CNT and graphene are mixed in either ethanol or acetone. Weighed quantities of graphene and CNT are added into ethanol or acetone and the beaker is inserted into a water bath or ultrasonic agitation for a definite period. This method provides broader chances for uniform mixing of the substrates. Yen and co-workers put forward a two-step solution-based method at room temperature (RT) for the preparation of graphene–MWCNT hybrid materials. The method involves the synthesis of graphene oxide from graphite by Staudenmaier's method followed by thermal reduction to graphene at a temperature of 1050°C. After that, graphene will be mixed with acid-treated MWCNTs and ultrasonication will be used to create the final hybrid. This will help to avoid the agglomeration of graphene sheets. The steps in the method are illustrated in Figure 1.2.[15]

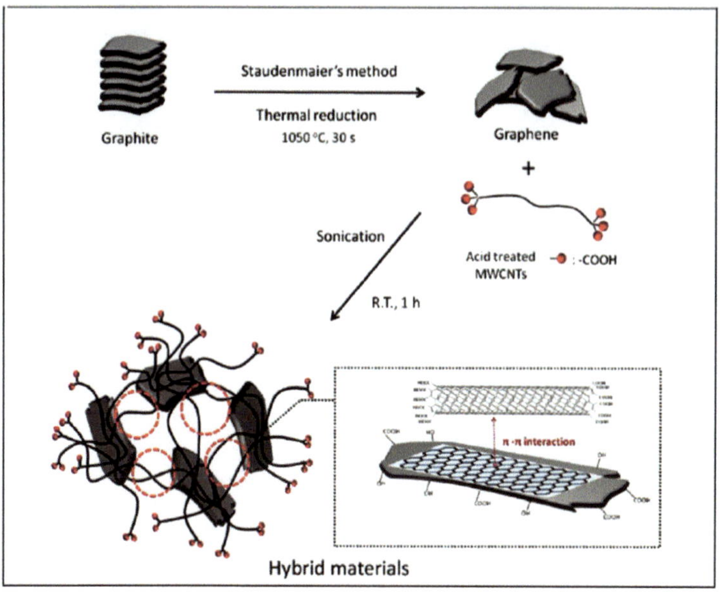

FIGURE 1.2 The schematic mechanism for the preparation of graphene–CNT hybrid material (Reproduced with permission from Ref. [15], Copyright 2014, Elsevier Publications).

Arc-discharge method is another important method for the in situ synthesis of graphene/SWCNT hybrid. The hybrid was prepared where the SWCNTs were grown in between or around the graphene sheets. This will form a network. Consumable anodes are created using the arc-discharge process by combining a catalyst, such as a combination of NiO and Y_2O_3, with a carbon source. The direct current arc discharge is carried out in a water-cooled stainless steel chamber filled with gases, such as H_2 and He at a suitable pressure. The discharge voltage is kept around 30 V by controlling the distance between the electrodes. By this method, large amounts of graphene/CNT hybrids can be synthesized in situ. Catalyst is removed from the reaction system by refluxing the hybrid in nitric acid. Hybrid is then neutralized by washing with deionized water till pH reaches 7. Prepared materials can be used in supercapacitors. In the manufacture of supercapacitors, first, the graphene/CNT hybrids are activated by KOH and are heated. Activated graphene/CNT hybrids are then used for making electrodes. Figure 1.3 illustrates the formation of graphene/CNT hybrids.[16]

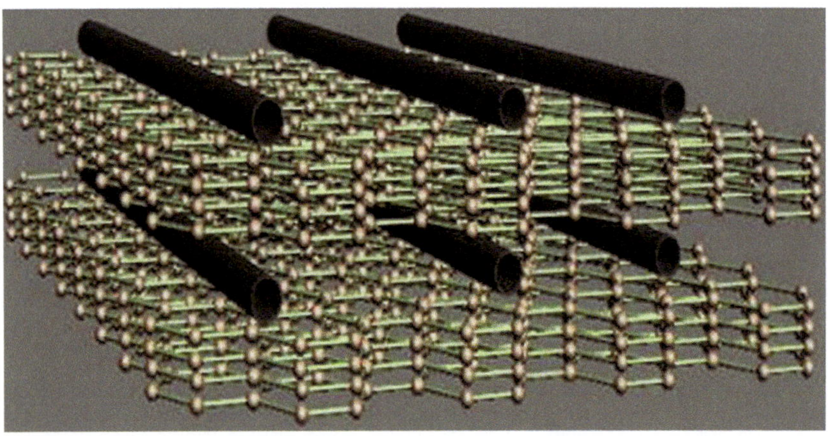

FIGURE 1.3 The hybrid materials are depicted with SWCNTs grown in between the graphene sheets, forming a unique network carbon nanomaterial, with both SWCNTs and graphene generated in situ and with fewer flaws than other approaches. (Reproduced with permission from Ref. [16]© 2012 Elsevier.)

Aqueous gel precursors treated by supercritical CO_2 drying can also be used to make ultralight graphene–CNT hybrid aerogels. Graphene oxide needed for the synthesis can be obtained by conventional methods. CNTs can also be obtained by the common synthetic methods. Predispersed graphene oxide sheets and CNTs are mixed thoroughly and sonicated to get homogeneous hybrids. Figure 1.4 schematically represent the green synthesis of graphene–CNT hybrid aerogels.[17]

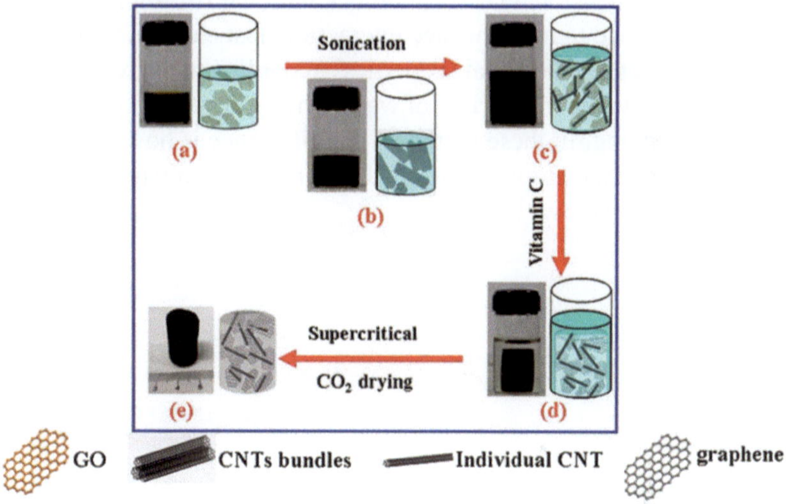

FIGURE 1.4 The green production of graphene–CNT hybrid aerogels: schematic diagram. (Reproduced with permission from Ref. [17]. ©1991. Royal Chemical Society.

Graphene–CNT hybrids can also be synthesized through self-assembly processes. Usually, it requires a metal substrate. Huang et al. prepared a graphene–CNT hybrid on a titanium substrate by simple casting method. The process involves the preparation of graphene oxide by Hummer's method and MWCNTs by the CVD method. Prepared graphene oxide and MWCNTs are then mixed by ultrasonication followed by self-assembly on titanium substrate to get the hybrid. Studies on vacuum-assisted self-assembly have also been carried out by researchers.

1.3 CARBON NANOTUBE–GRAPHENE HYBRIDS AN OVERVIEW

As mentioned in the previous section, the hybrids are meant to improve the properties of materials. When considering the nanomaterials, a large number of hybrids are possible and among them, CNTs and their hybrids play an inevitable role. Among the pi bonded nanofillers, CNTs, and graphene are the most favorite candidates and when these two are combined to form a hybrid, their properties become incomparable. The idea behind the fabrication of hybrids from these materials is due to the high mechanical and electrical properties of both materials. The combinations of one-dimensional (1D) nanotubes with the two-dimensional (2D) materials will lead to more surface interaction, fast electron transfer which in turn enhances the electrical properties. According to the reports, 1D–2D pairings can develop in a variety of ways. They can be adsorption, penetration and wrapping. In these methods, the difference is how the nanotubes are connected to the graphene sheets. It can be pictorially represented as in the image given below (Fig. 1.5).

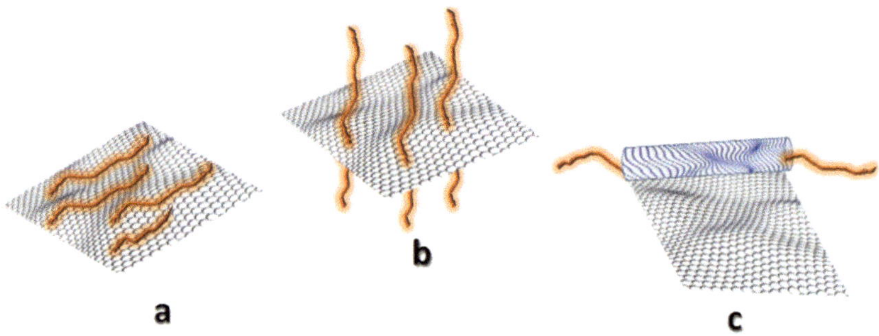

FIGURE 1.5 Different modes of interaction between graphene and CNTs: (a) adsorption, (b) penetration, and (c) wrapping.

The type of interaction and properties will depend on the contact site of the CNTs with graphene. CNTs, graphene, their functionalized versions, and graphene oxide sheets can all be combined.

1.3.1 METHODOLOGIES ADOPTED IN THE FABRICATION OF HYBRIDS

General strategies of the hybridization between graphene and CNTs can be categorized into two.

(a) The assembly method and
(b) In situ methods

1.3.1.1 THE ASSEMBLY METHOD

This method involves the self-assembly and chemical assembly of the graphene sheets with nanotubes. In the assembly method, the π systems (sp^2 hybridized orbitals) will interact together to form hierarchical hybrids. Pristine materials as well as functionalized ones can easily form the assembly because of the π-π interaction of the network of C=C. Functionalization on the surfaces of anyone or both will enhance the possibilities of covalent linking, π-π stacking or van der Waal type of interactions. The assembly method resulted in a high surface area and three-dimensional network of CNT and graphene. But one of the severe disadvantages of this method include (1) the process is a multistep process and (2) there is difficulty in controlling the number of layers. Assembly to form hybrids are possible by many techniques, such as solution processing, layer-by-layer assembly, vacuum filtration.

(i) Solution Processing

The liquid-based processing technologies are favorable to the industrial synthesis of nanomaterials. Kim et al reported the self-assembled synthesis of graphene oxide–CNT hybrid films on a Ti template using aqueous dispersions. An intercalated structure of CNTs between stacked graphene oxide sheets was obtained which processed enhanced electrochemical properties. Yang et al. reported the synthesis of transparent hybrids from CNT and GO. They have presented a novel route using anhydrous hydrazine. In the method, the GO was synthesized using the Hummers method and the dry powdered GO was coupled with CNTs functionalized with acid carboxy and hydroxyl functionalities. Both of them was directly dispersed in anhydrous hydrazine and a range of composition was prepared. The suspension was found stable for months. They used the materials for the manufacturing of polymer-based solar cells and had good power conversion efficiency.[18] A representative SEM image of the hybrid is given in Figure 1.6.

Cai et al. also used the Hummers method for the synthesis of GO and fabricated the hybrids with CNTs using dispersions of DMF. The fabricated films presented multilayered morphology. They were suitable for electrical applications and the conductive properties enhanced with the added CNT content.[19] Single-walled CNT/GO hybrids were synthesized using solution processing by Huang et al. In their report, they used different doping alkali carbonate salts. The hybrids were suitable for photovoltaics.[20] A method for making integrated CNT/graphene hybrids using water was presented by

Song et al. The process is detailed as first the CNTs were mixed with the GO solution, followed by ultrasound sonication for hours. The resulting mixture was dried and reduced in an H_2 atmosphere and another method CNTs and reduced GO were also directly mixed by ball milling. The two methods were compared for efficiency and the schematic representation is given in Figure 1.7. Both the CNT/graphene hybrid materials were employed as cathode materials and were assembled into field emission devices (FED).[21]

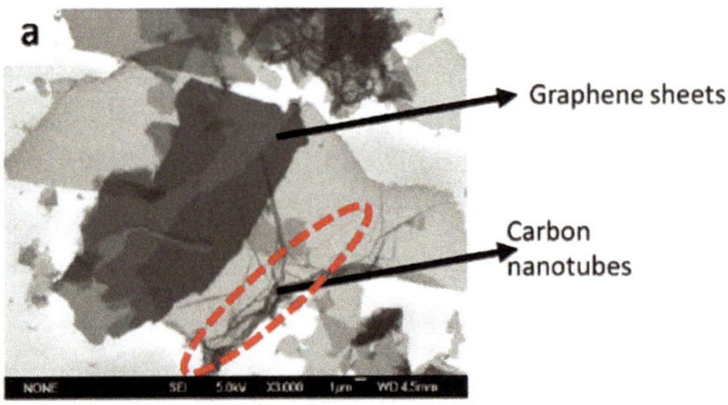

FIGURE 1.6 Scanning electron microscope representation of CNT–graphene hybrids.
Source: Reproduced with permission from Ref. [18]. © 2009 American Chemical Society.

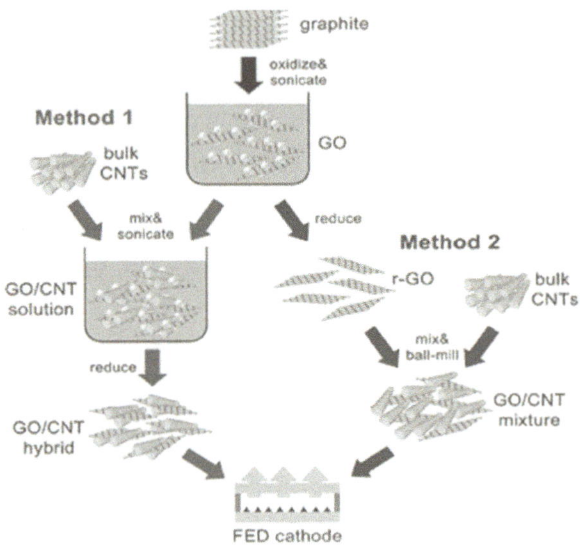

FIGURE 1.7 Comparison of methods of hybrid formation.

A water-based method was also reported by King et al. in which a small amount of graphene was used to prepare hybrids and to improve the electrical conductivity of SWCNT films. The hybrid materials reported that CNTs–GO hybrid thin films possessed excellent electrochemical performance as electrodes for supercapacitors.[22] Hydrazine monochloride-DMF solution-mediated functionalized CNT and reduced GO nanosheets were fabricated for laser desorption/ionization mass spectrometry of small molecules and tissue imaging by Kim and co-workers.[23] Song and co-workers used the solution processing method for hybridization. An aqueous solution dispersion of positively charged horseradish peroxide (HRP) was introduced into the GO–MWNTs at the surface of the glassy carbon (GC) electrode to form an HRP/GO–MWNT/GC electrode under mild conditions. The hybrids showed electrocatalytic activity.

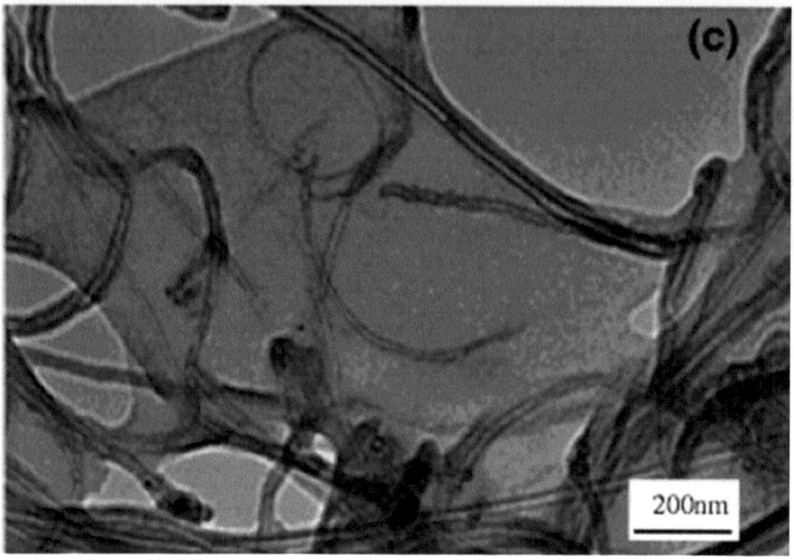

FIGURE 1.8 TEM images of MWNT–GO hybrids (Reproduced with permission, Copyright 2013, Springer).

The morphology analyzed by TEM revealed the bundles of MWNTs adsorbed on the surface of the nanosheet to form an integrated nanocomposite (Fig. 1.8). With the help of water-soluble GO in the composite, GO–MWNTs was well-dispersed in an aqueous solution.

(ii) Layer-by-Layer Assembly

Layer-by-layer assembly method is used to fabricate multilayer materials. The layer-by-layer method is also represented as the chemical method in which the functionalized materials are coupled by wetting with solvents (Fig. 1.9). The key advantages of this method are that the procedure is simple, they offer uniform dispersion, controlled multilayers, etc. The layer-by-layer assembly formed through electrostatic interactions was reported by Hong et al. In their report oppositely charged suspensions of the RGO nanosheet electrostatically interact with MWNTs. This method produced hybrids with excellent optical and conductive properties. The conducting network of MWNTs provided additional flexibility and mechanical stability for GO nanosheets and these hybrids were found suitable for the application as a highly flexible and transparent electrode.[24] In their work, Kim and Min fabricated the CNT–graphene hybrids via self-assembly. The graphene produced by Hummers method and CNT synthesized by chemical vapor deposition method was combined and fabricated on a silicon substrate treated with a toluene solution of 3-aminopropyl triethoxy silane. The aminated CNTs were incorporated in the sheets of graphene and they obtained transparent hybrids.[11]

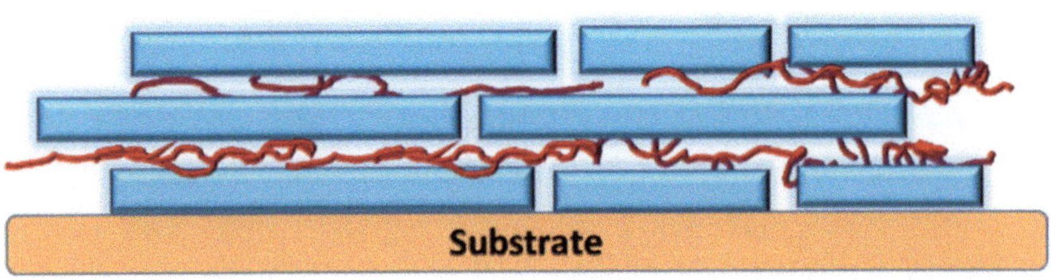

FIGURE 1.9 The schematic representation of the hybrids formed by layer-by-layer assembly.

Yu and Dai reported the hybridization by self-assembly by the L-b-L method. In their report, they chemically reduced graphene oxide nanosheets with hydrazine in the presence of poly (ethyleneimine) as a stabilizer. This introduced a charged soluble polymer unit onto the graphene and resulted in a well-dispersed graphene-based material.[25] It is represented in Figure 1.10

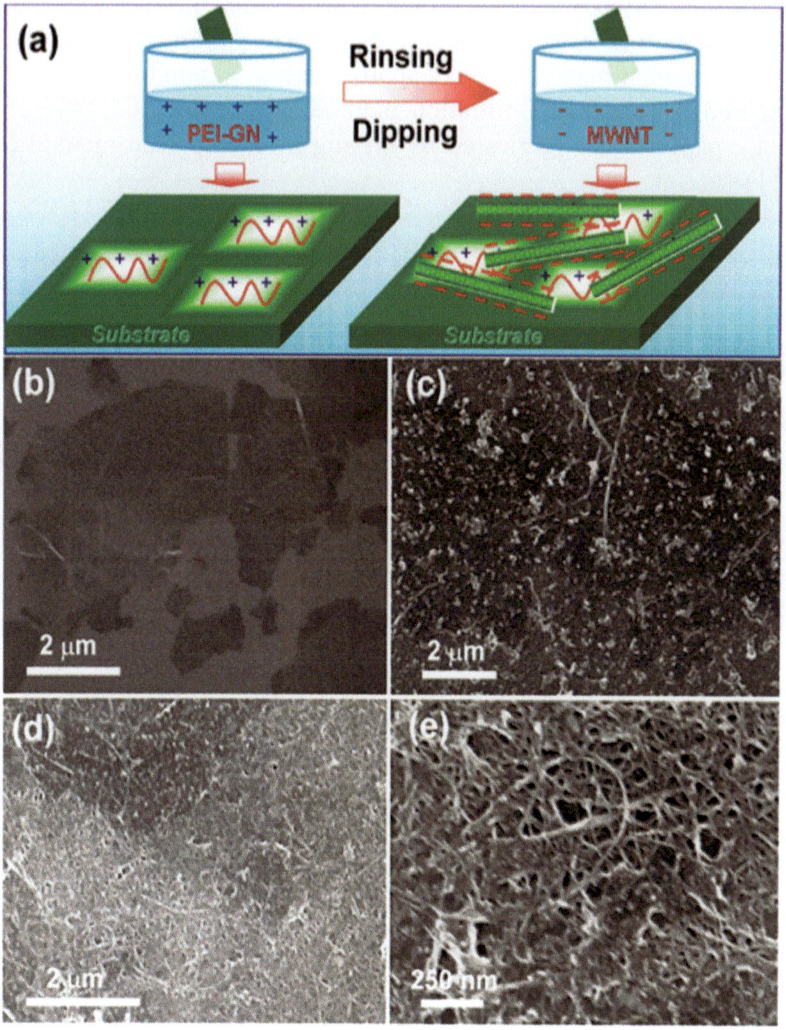

FIGURE 1.10 Schematic representation of the graphene-functionalized CNT hybrid (a) and the SEM images (b) (Reproduced with permission from Ref. [25], Copyright 2010 American Chemical Society).

(iii) Vacuum Filtration

Low cost, being scalable, and low-temperature preparation are the main advantages of the vacuum filtration methods. Feng et al. used this method for the fabrication of hybrids for the field emission devices. The vacuum filtration method was reported to synthesize carbon nanotube bucky papers. Circular bucky papers of approximately 33 mm in diameter were prepared by a vacuum filtration method. The hybrid CNT/Graphene bucky papers were fabricated by mixing a co-dispersed method (Fig. 1.11).

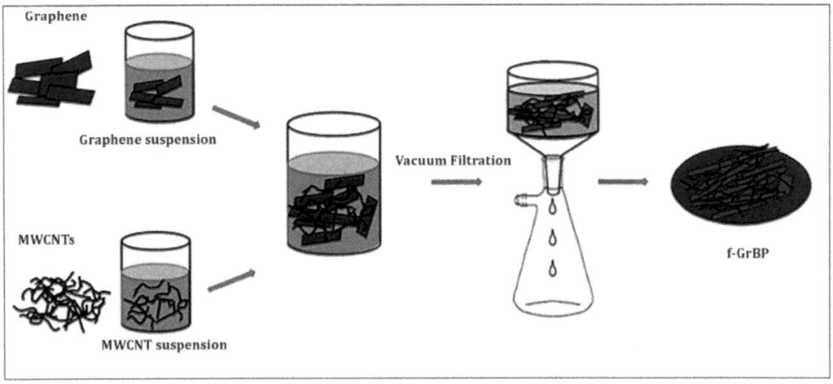

FIGURE 1.11 Schematic representation of the preparation of bucky papers of CNT/graphene hybrid.

Initially, two separate suspensions of 20 mg nanotubes and 5, 10, or 20 mg of graphene in propanol were prepared by bath sonication and magnetically stirred overnight. Then the suspensions were mixed and stirred again for 6 hours. The resulting bucky papers were obtained after the vacuum filtration and drying process.[26] Porous graphene–CNT hybrid nanopaper was prepared by Liu et al., and in their article, they have detailed the synthesis and the applications. For the fabrication, the aqueous colloidal suspensions of graphene oxide sheets were mixed with CNT accumulations and ultrasonicated for an hour. The obtained materials were treated with polystyrene suspension. The hybrids of CNT–graphene containing polystyrene spheres were formed. GO– CNT hybrids were then vacuum-filtered through a poly(vinylidene fluoride) (PVDF) membrane filter. They converted the hybrids to paper by treating with polyaniline particle and the resultant ternary papers were analyzed for electrochemical properties.[27]

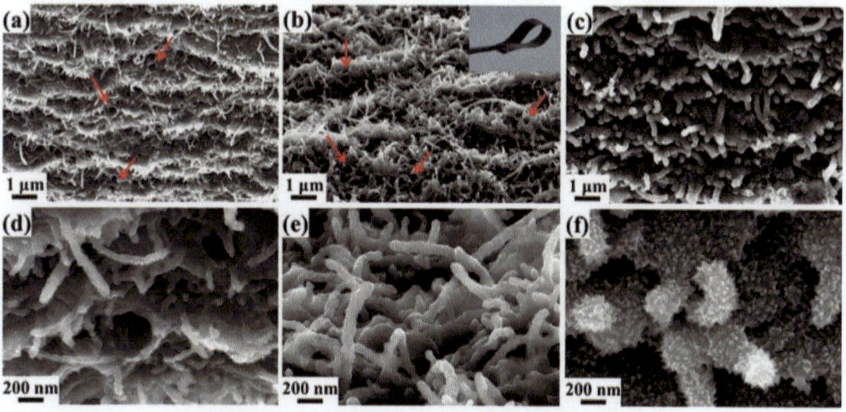

FIGURE 1.12 Cross-sectional FESEM images of (a and d) p-GC/PANI1, (b and e) p-GC/PANI3, and (c and f) p-GC/PANI5 papers at (a–c) low and (b–f) high magnifications (Reproduced with permission from Ref. [27], Copyright 2011 RSC).

1.3.1.2 IN SITU METHODS

The reduction in the quality of the hybrids by the assembly method can be overcome by the in situ methods. In this category, the two techniques involved are chemical vapor deposition and unzipping.

Chemical Vapor Deposition (CVD)

This is a common method adopted for the synthesis of CNTs. The hybridization and the dimensions can be controlled by controlling the conditions of CVD. Uniform dispersion is ensured in this technique. In an interesting work, Ramaprabhu et al. reported the wrapping of CNTs by graphene sheets to fabricate the hybrids (Fig. 1.13). They have analyzed the morphology of the wrapped hybrids and found that the graphene sheets are wrapped over the length of carbon nanotube like a smooth coating. The preparation method involved one-step chemical vapor deposition growth on a catalytic mixture of Misch metal-based alloy hydride ($MmNi_3$) and graphene oxide. The hybrid composite possessed the advantage of having strong interaction between CNT and graphene rising from an in situ preparation technique.[28]

Microwave plasma chemical vapor deposition method for the graphene–CNT hybrids for the field emission application was reported by Vanker et al. They have presented an aligned CNTs network on the structure and above this, a uniform distribution of the graphene sheets is observed. The layer thickness can be tuned by deposition methods. When the time of deposition growth time increases, the density of the layered graphene decreases and the size increases.[29]

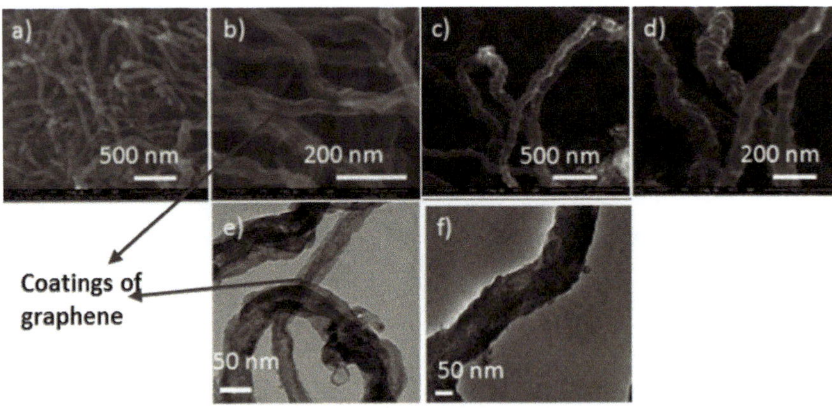

FIGURE 1.13 The FESEM images of graphene wrapped CNTs at different magnifications (Reproduced with permission from Ref. [28], Copyright 1991 RSC).

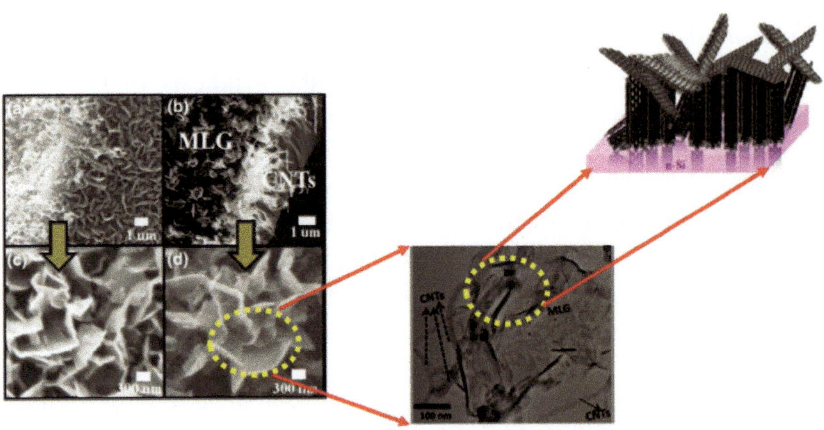

FIGURE 1.14 The graphene–CNT hybrids obtained by microwave plasma-assisted chemical vapor deposition (Reproduced with permission from Ref. [29], Copyright 2014 Springer).

1.3.2 CHARACTERIZATION TECHNIQUES

The characterization techniques of the CNT–graphene hybrids and their functionalized derivatives depends on the end application of the materials However, the general characterization mainly involves the analysis of morphology using different microscopic techniques, such as scanning electron microscope, transmission electron microscopy, and atomic force microscopy. The hybrid formation or inclusion is confirmed from these techniques. Dong and co-workers reported the fabrication of hybrids with enhanced thermal and electrochemical properties using stainless steel substrates.[30] The SEM images of the pristine CNTs and CNT–Graphene hybrids presented interesting morphologies. The graphene sheets and the bundles of CNTs are very clear in these images. The cross-section of the hybrids represented by the image f shows that the carbon nanotube surface was fully covered by the graphene sheets (Fig. 1.15).

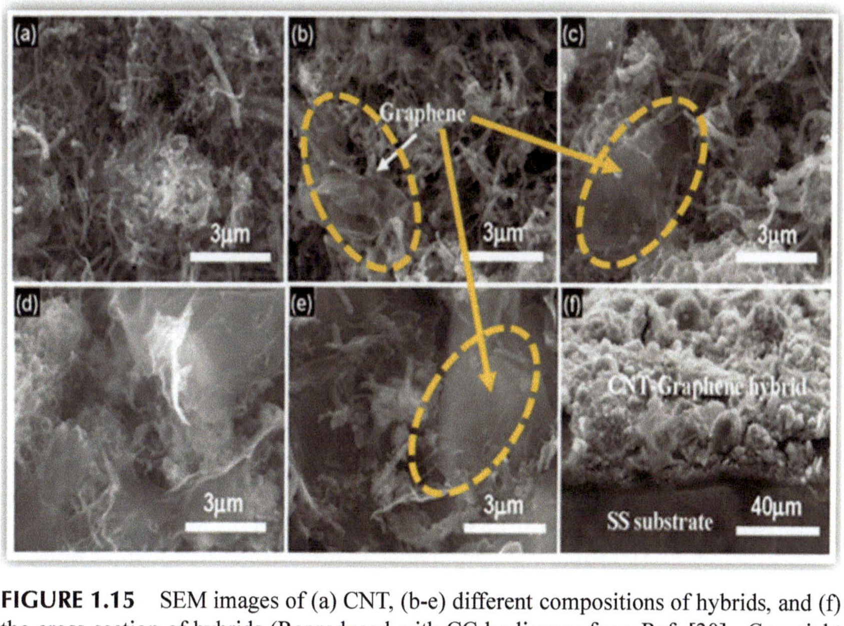

FIGURE 1.15 SEM images of (a) CNT, (b-e) different compositions of hybrids, and (f) the cross-section of hybrids (Reproduced with CC by license from Ref. [30], Copyright 2019 MDPI).

The effect of growing time in the CVD method can be well understood by the FESEM images. Lin and Lin studied the synthesis and application of CNT/graphene hybrids for electrochemical capacitors. The carbon nanotube/graphene composites grown by one-step CVD for 10 min is a little smaller than others (0.448 and 0.493) since the packing density of CNTs (possessing more defects than graphene) suddenly decreases with the decreasing growing time (Fig. 1.16).[31]

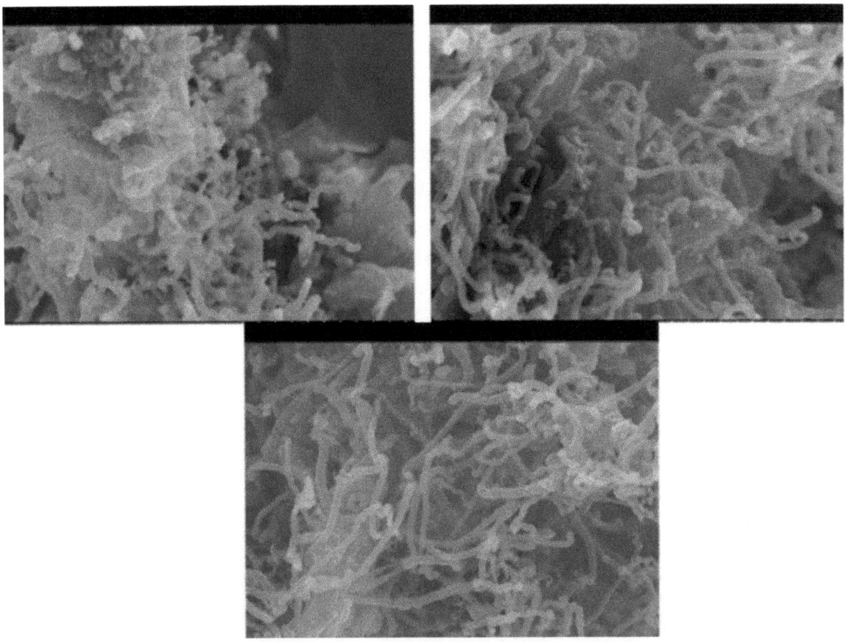

FIGURE 1.16 The FESEM images of carbon nanotube/graphene composites grown by one-step CVD at 850°C for different growing times (Reproduced with permission from Ref. [31], Copyright CC by licence (2019)). https://creativecommons.org/licenses/by/4.0/

Electrical characterizations are important for field emission applications and electronic applications. To achieve improved FE characteristics, it is ideal that the material has good conductivity as well as possessing areas of local field enhancement. These can be in the form of (i) physical surface protrusions or (ii) areas where field lines are concentrated at the surface of the emitter, due to different conductivities of phases in a composite. The current–voltage relationship is a characteristic technique used in electronic applications. Chua and co-workers investigated the effectiveness of the CNT/graphene hybrids prepared by electrophoretic deposition. They found that the turn-on time decreased with increasing the deposition time.[32]

The crystallinity and purity can be analyzed using X-Ray diffraction studies and Raman scattering studies. Muangrat and co-workers studied the graphene nanosheet-grafted double-walled carbon nanotube hybrid nanostructures by two-step chemical vapor deposition and they obtained four significant Raman peaks generated from the hybrids, which include. DWCNTs-derived radial breathing mode (RBM) around ~150–27 cm^{-1}, disorder carbon-derived D– at ~134 cm^{-1}, graphitic structure-derived G-band at ~159 cm^{-1}, and second-order of D-band-derived 2D-band at ~267 cm^{-1}. In addition, the spectrum of G(20)-DWCNTs shows the D-band at 1620 cm^{-1}, which originates from the intravalley double-resonance process.[33]

1.4 SIGNIFICANCE OF CARBON NANOTUBE-GRAPHENE HYBRID FOR ENERGY STORAGE

Graphene–CNT hybrids found large applications in supercapacitors. The hybrid material shows a good synergistic effect which will give higher capacitance compared with graphene and CNTs alone. A maximum specific capacitance of 251 F/g was achieved at 5 mV/s in 1000 cycles for the hybrid material which is considerably higher than that of the individual CNT (85 F/g) or graphene oxide (60 F/g). Graphene–MWCNTs show exceptionally good electrocatalytic redox reversibility in flow batteries. The Ipa/Ipc value remains constant even at a higher scan rate. A significant decrease in charge transfer resistance across the electrode/electrolyte interface has occurred. Graphene–CNT hybrids can be effectively used as an anode modifier for polymer solar cells. The hybrid offered better performance compared with the conventional anode modifier. The presence of CNTs in the graphene sheets helps to promote the hole extraction and charge flow in the anode modifying layer. This will offer higher optical transmission in the longer wavelength regions.[15]

1.4.1 USE OF HYBRIDS IN FUEL CELLS

The combination of one-dimensional carbon nanotube (CNT) and two-dimensional graphene has exceptional advantages when used as a catalyst support material in fuel cells. Ultrathin graphene–multiwalled carbon nanotube hybrid material produced by a solar exfoliation technique has been explored as catalyst supports for oxygen reduction reaction (ORR) in proton exchange membrane fuel cells (PEMFC).[34] Pt nanoparticles distributed over a solar exfoliated graphene-functionalized multiwalled carbon nanotube (sG-f MWNT) hybrid nanocomposite exhibit higher ORR electrocatalytic activity than Pt dispersed functionalized solar graphene (f-sG) catalyst support. As synthesized graphene is hydrophobic because of the removal of functional groups over the basal planes of GO, and hence has to be surface modified to anchor Pt nanoparticles. But the usual surface functionalization strategies of graphene consequences in restacking of the graphene sheets thus decreases the effective surface area. This can be avoided by using spacers, and CNTs can be employed as space impediments between the graphene sheets to prevent the restacking of graphene. The maximum power density of the Pt/sG-f MWNT is 675 mW cm^{-2}, which is 1.9 times higher than Pt/f-sG (355 mWcm^{-2}). The enhanced performance of the sG-f MWNT composite as catalyst support compared with pure f-sG has been ascribed to the bridging of defects for the electron transfer and a rise in the basal spacing between graphene sheets with the addition of MWNT. In another study of graphene nanosheet, CNT hybrid nanostructure is used as an electrode for a proton exchange membrane fuel cell. This system attained fuel cell performance of the maximum power density of 1072 mW cm^{-2} at 80°C under H_2/O_2. These enhanced performances can be credited to a great surface area, homogeneously dispersed Pt nanoparticles, outstanding electrical conductivity, and a direct electron-transfer pathway of the GNS-CNT electrode.[35]

Hierarchically porous MWCNT@rGO hybrids are fabricated by Zou et al. to accelerate the extracellular electron transfer of anodic biofilm in microbial fuel cells. Here, CNT plays a dual role, preventing the agglomeration of graphene and also acts as connections to upsurge multidirectional contacts between 2-D rGO sheets.[36] Wu et al. created a hierarchically porous nitrogen-doped CNTs/reduced graphene oxide (rGO) composite using polyaniline as a nitrogen source for microbial fuel cells anode. The maximum power density achieved with the N-CNTs/rGO anode in *Shewanella putrefaciens* CN32 MFCs is 1137 mW m^{-2}, which is 8.9 times compared with carbon cloth anode and also higher than that of N-CNTs (731.17 mW m^{-2}), N-rGO (442.26 mW m^{-2}), and CNTs/rGO (779.9 mW m^{-2}) composite without nitrogen doping. The greatly improved bioelectrocatalysis could be attributed to the enhanced adsorption of flavins on the N-doped surface and the high density of biofilm adhesion for fast interfacial electron transfer.[37] Prasad et al. fabricated three-dimensional graphene-carbon nanotube hybrid for high-performance enzymatic biofuel cells. Here, 3D graphene provides a large surface area for loading enzymes and their catalytic reactions. The synergistic effect of 1D CNT and 2D graphene promotes charge transfer and conduction. The fabricated biofuel cell has got an open circuit voltage (1.2 V) and a high power density (2.27 ± 0.11 mW cm^{-2}) (Fig. 1.17).[38]

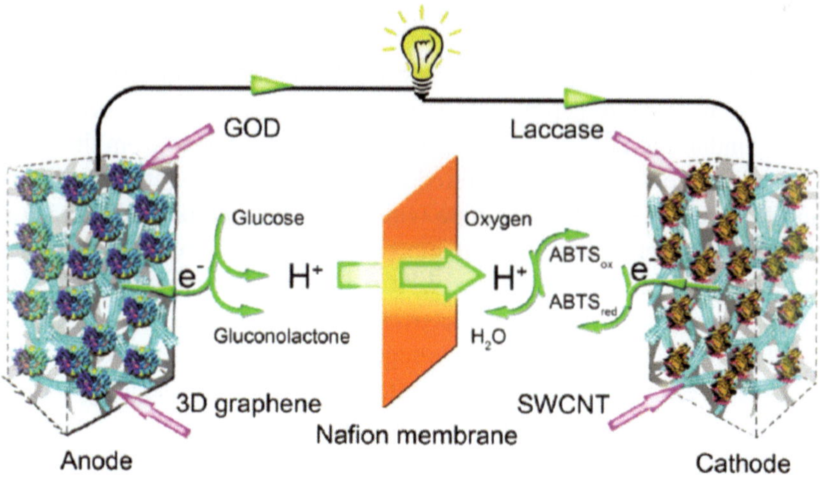

FIGURE 1.17 Illustration of the EBFC equipped with 3D graphene–SWCNT hybrid electrodes (Reproduced with permission from Ref. [38], Copyright 2014 ACS).

1.4.2 CNT-GRAPHENE HYBRIDS FOR SUPERCAPACITORS

The energy storage devices that store and distribute electrical energy as a result of a reversible adsorption–desorption process of the ions at the interface that exists between the electrolyte and electrode materials are considered supercapacitors. They are having very high capacitance than the normal capacitors and exceptional charge/discharge cycle and hence they are often referred to as ultracapacitors. In electronic systems, supercapacitors are used in applications such as power supply stabilizers, power supply pulses, grid power buffer and for other energy storage and energy recovery. Based on the charge storage mechanism, supercapacitors are divided into pseudocapacitors, electrical double-layer capacitors, and hybrid super-capacitor. Many materials are being employed in the supercapacitors. Among these materials, carbon-based materials are the favorite candidates. Owing to the superior electrical properties of carbon-based materials, especially the ones in the nanometer scale, such as CNTs, graphene,. they possess high conductivity, lightweight nature, and electrochemical stability.[39]

When the carbon-based materials are coupled together to form hybrids, the properties are proportionally enhanced. In most of the capacitors, nanotubes either single or multiwalled are hybridized with graphene and its surface modifications. The duo of CNTs and graphene for supercapacitor applications has been extensive. The CNT-graphene combination can be from the in situ interaction, self-assembly, or by coupling or combination with conducting polymers. The graphene sheets obtained from the reduction of graphene oxide are modified with a polymer, poly(ethylene imide), and then it is allowed to combine with acid functionalized CNTs to form the hybrid sheets. These combination films were shown to have an interconnection of carbon structures with well-defined nanopores which are promising for supercapacitor electrodes, exhibiting a nearly rectangular cyclic voltammogram with an average specific capacitance of 120 F/gg even at an exceptionally high scan rate of 1 V/s.[26] The current generation of electronic material prefers lightweight, thin, and flexible materials in their applications. Paper capacitors offer these characteristics and have thus been developed mostly. Bonaccorso and co-workers for the realization of high-performance supercapacitors, suggested a graphene/single-walled carbon nanotube (SWCNT) hybrid structure. The devices are constructed using 2 cm^2 hybrid graphene/SWCNT electrodes, designed by integrating complex spray-gun deposition and an aqueous electrolyte with a scalable solution processing system. This design enabled the output of supercapacitors to be efficiently balanced by achieving remarkable electrode gravimetric capacitance and specific energy at 2 mV s^{-1} of 104 F g^{-1}, 20.8 Wh kg^{-1} and maximum specific strength of 92.3 kW kg^{-1}1.[40] Supercapacitor performance can be enhanced by the application of hybrid composites as electrode materials. Potentiostatic electrochemical polymerization method can be implemented to fabricate such electrode materials. In the process, the graphene and CNT are embedded simultaneously while pyrrole is being synthesized, forming a nanocomposite electrode film. As the pyrrole is polymerized, the anions are incorporated from the reaction solution, maintaining the charge neutrality in the polymer matrix. The composite electrodes when placed in the presence of sulfuric acid electrolyte yielded the highest specific capacitance, and exhibited an interesting morphology due to the well-defined crystal structure.[41]

CNT-graphene hybrid formation has a major advantage in that the agglomeration of CNT can be minimized and the restacking of graphene layers can be minimized. This is possible when there is a chemical bonding between them. Such a hybrid material can be easily employed in supercapacitors. Owing to the enlarged interlayer spacing up to 0.55 nm, the GO/CNT supercapacitor generated a volumetric capacitance of 165 F cm^{-3}, which is a performance much superior to other carbon-based electrodes.[42] Similar to this densely packed graphene nanomesh–carbon nanotube (CNT) hybrid films can be prepared through a simple graphene etching process and by subsequent vacuum-assisted filtration method. Due to the incorporation of CNTs between sheets and graphene nanomesh, the special arrangement endows rapid transport of electrolyte ions and electrons in the film electrode. As a result, the GNCN film electrode-based supercapacitor exhibits both high gravimetric and volumetric capacitances, as well as high volumetric capacitances. It can be used for applications, such as Li-ion batteries, storage of hydrogen, sensors and actuators, and purification of water.[43] Sandwiched graphene/CNT hybrids are employed in the supercapacitors as electrode materials. The integration of CNT into graphene by CVD or physical mixing will not only reinforce the graphene material surface region but also serves as a "spacer" to provide the diffusion path, promoting the rapid transport of electrolyte ions inside the electrode material, culminating in the improvement of CNT/graphene electrochemical properties. The specific structure provides easy transport of electrolyte ions or electrons within the electrode and full utilization of quasi- and double-layer capacitance (Fig. 1.18).[44]

A novel thought in the fabrication of the electrode materials is to introduce a porous electrode from natural materials. Cellulose is a better choice in these cases due to the wide variety of sources of cellulose. Cellulose nanofiber can be used to embed the CNT/graphene hybrid materials and the resulting materials were converted to aerogels. The CNF/RGO/CNT aerogel film can be used to fabricate the supercapacitors without the use of any other binders, current collectors, or electroactive additives. Symmetric supercapacitors were assembled using a thin, flexible polyethylene terephthalate (PET) substrate and H_2SO_4/PVA gel as the electrolyte and separator. The electrode exhibited a specific capacitance, energy density, and power density of 252 F g^{-1} (216 mF cm^{-2}), 8.1 mW h g^{-1} (28.4 μW h cm^{-2}), and 2.7 W g^{-1} (9.5 mW cm^{-2}), respectively. Moreover, the supercapacitors showed excellent cyclic stability, with more than 99.5% specific capacitance retained after 1000 charge–discharge cycles. This approach is significant in terms of low environmental impacts and low cost.[45] The supercapacitor materials can also be fabricated with the help of transition metal compounds and other inorganic or inorganic materials, such as MoS_2, metal oxides etc.[46,47]

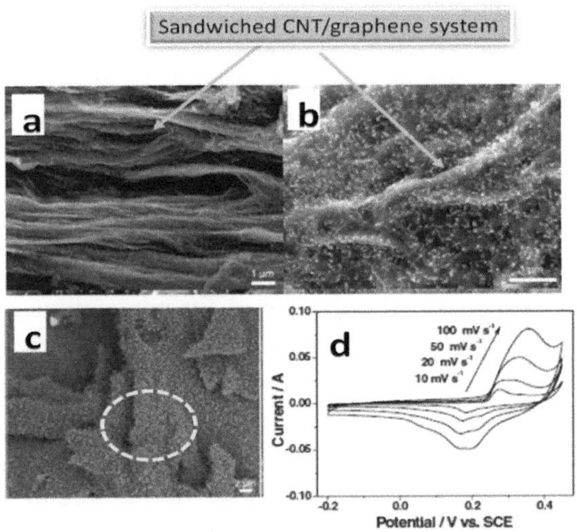

FIGURE 1.18 The SEM images of the sandwiched structure of CNT/graphene hybrid sheets and the CNT insertion is highlighted with the circle. d represents the electrochemical performance (Reproduced with permission from Ref. [44], Copyright 2010, Wiley).

1.4.3 CNT-GRAPHENE HYBRIDS FOR SOLAR CELLS

Hybrids consisting of CNTs (SWNTs) and graphene heterostructures bonded covalently or by weak van der Waals forces have been developed and performs superior properties compared with individual components. Graphene–SWNT assemblies exhibit astounding properties of mechanical flexibility and stretchability as well as electrical and optical conductivity.

This will enable them to give applications in optoelectronics. These devices exhibited high photoresponsivity and fast response time upon visible irradiation. The fabricated materials show flexibility and good folding strength. Therefore, these materials show potent practical applications in the field of photosensors, flexible solar cells, etc. [48]

Graphene and CNTs hybrids are having several superior properties over conventional conducting electrodes used in solar cells made up of CdTe, $CuInS_2$ chalcogenides. CNT–graphene-based nanolayers possess high transparency for a wide range of wavelengths, high mechanical and thermal stability, very low reflectance, and hole-selective conductivity. The cost of production is also low when compared with conventionally used layers. Nevertheless, a few researchers have applied nanolayers in the structure of TFPVs. Application of graphene–CNT hybrids can be seen in DSSCs, Perovskite, and semi-transparent solar cells.[49]

1.4.3.1 CNT–GRAPHENE HYBRIDS IN PEROVSKITE SOLAR CELLS

Various materials can be used to improve the device stability and efficiency of perovskite solar cells which include interfacial materials, such as metal nanostructures, carbon materials, water-resistant materials. They modify the interface between the perovskite and hole-transport layers. Wong and co-workers reported that CNTs can act as an efficient hole collector in solar cells due to their capacity to extract more charge and the probability of recombination of CNT hole is also low. To this, graphene oxide and reduced form of graphene oxide can be employed as hole conductors to synthesize efficient perovskite solar cells. CNTs with large diameters are generally used in solar cells because CNTs with smaller diameters cause difficulty to get dispersed on the layers. CNTs with large diameters are less expensive also.[50] CNT–graphene hybrids can be synthesized by growing multilayer graphene on CNT cores using ethanol as a precursor by the plasma-enhanced CVD method at optimum conditions. Then a solution mixture of CNT–graphene hybrid and Spiro was spin-coated on top of the perovskite film of the solar cell to get more efficient.[51] Studies have shown that a simple perovskite solar cell shows an efficiency of about 15.67%, whereas CNT–graphene-doped perovskite solar cells show increased efficiency of about 18.12%. Photoluminescence (PL) characterization was carried out to understand the carrier extraction efficiency of PSC devices.

In CNT–graphene hybrid incorporated solar cells, the PL intensity of the film decreases. This ensures the perfect hole extraction in the film. Transient photocurrent (TPC) and voltage (TPV) measurements were carried to determine the hole transport and recombination time constants. The CNT–graphene-based devices show more rapid photocurrent decay. This indicates the faster hole transport in the solar cells. The slower photovoltage decay confirms the improved retardation of carrier recombination compared with that in the pure perovskite solar cells. The extraordinary bonding between graphene and the CNTs contributes toward the quick charge transport by graphene to the CNTs which will lead to the enhanced charge transport of the devices.[52] Another important property of these devices is their high stability in humid environments. CNT–graphene incorporated perovskite solar cells show tremendous thermal stability under thermal annealing.[3]

1.4.3.2 CNT-GRAPHENE HYBRIDS IN DYE-SENSITIZED SOLAR CELLS

It is found that CNTs bridged with graphene nanoribbons can be used as better replacements for conventional platinum counter electrodes in DSSCs. The energy conversion efficiency of a conventional DSSC was found to be 7.61% whereas in CNT–graphene hybrid incorporated DSSC shows approximately 8.23%. The CNTs were unzipped chemically to produce CNT–graphene nanoribbon hybrid. This was incorporated in the DSSC by the spin coating method. These can be used as the counter electrodes to fabricate DSSCs. Working electrodes have several choices. Fluorine-doped tin oxide with titanium dioxide nanoparticles is one among them. Dyes like cis-isothiocyanate-bis (2,2′-bipyridyl-4,4′-dicarboxylate) ruthenium (II) bis(tetrabutylammonium) can also be employed. It is seen that the CNT–graphene hybrid incorporated DSSC and the one without the hybrid gives almost the same open-circuit voltage. But they exhibited different short-circuit current density and filling factor values. In the bridged structure, electrons are rapidly transported along the bridges among CNTs from the inner side to the conductive substrate. CNT–graphene hybrids can be made transparent and flexible. This shows several applications in a wide variety of optoelectronic devices, such as solar cells and organic light-emitting diodes. Chemical composition, density, and alignment played critical roles in the resistance of hybrid films.[53]

1.4.3.3 CNT-GRAPHENE HYBRIDS IN SEMI-TRANSPARENT SOLAR CELLS

Semi-transparent solar cells (SSCs) show eye-catching appearances and great interest for their use of power-generating windows for buildings and vehicles, high-efficiency tandem solar cells and wearable electronics. SSCs can harvest a very large amount of sunlight from both sides. The incorporation of CNTs and graphene into SSCs can enhance their optoelectronic properties, such as high optical transparency, low sheet resistance, and high mobility. They also offer mechanical flexibility and environmental stability. The cost of production is comparatively low when compared with conventionally used electrodes. [54]

1.5 CONCLUSION

Hybrids of CNT and graphene are found to be excellent materials for energy storage applications. This is because of the synergism between these materials. This hybrid combination is useful in fuel cells, supercapacitors, and solar cells.

KEYWORDS

- **hybrids**
- **energy storage**
- **solar cells**
- **CNT**
- **graphene**

REFERENCES

1. Liu, C.; Li, F.; Ma, L.-P.; Cheng, H.-M. Advanced Materials for Energy Storage. *Adv. Mater.* **2010,** *22*, E28–E62.
2. Sahoo, N. G.; Pan, Y.; Li, L.; Chan, S. H. Graphene-Based Materials for Energy Conversion. *Adv. Mater.* **2012,** *24* (30), 4203–4210.

3. Zhu, J.; Yang, D.; Yin, Z.; –Yan, Q.; Zhang, H. Graphene and Graphene-Based Materials for Energy Storage Applications. *Small* **2014,** *10* (17), 3480–3498.
4. Ibrahim, H.; Ilinca, A.; Perron, J. Energy Storage Systems—Characteristics and Comparisons. *Renew. Sustain. Ener. Rev.* **2008,** *12*, 1221–1250.
5. Zhao, X.; Sanchez, B. M.; Dobson, P. J.; Grant, P. S. The Role of Nanomaterials in Redox-Based Supercapacitors for Next Generation Energy Storage Devices. *Nanoscale* **2011,** *3*, 839.
6. Guoa, S.; Wang, E. Noble Metal Nanomaterials: Controllable Synthesis and Application in Fuel Cells and Analytical Sensors. *Nano Today* **2011,** *6*, 240–264.
7. Zhai, Y.; Dou, Y.; Zhao, D.; Fulvio, P. F.; Mayes, R. T.; Dai, S. Carbon Materials for Chemical Capacitive Energy Storage. *Adv. Mater.* **2011,** *23*, 4828–4850.
8. Wang, B.; Hu, C.; Dai, L. Functionalized Carbon Nanotubes and Graphene-Based Materials for Energy Storage. *Chem. Commun.* **2016,** *52*, 14350.
9. Yamashita, S.; Martinez, A.; Xu, B. Short Pulse Fiber Lasers Mode-Locked by Carbon Nanotubes and Graphene. *Opt. Fiber Technol.* **2014,** *20* (6), 702–713.
10. Zhu, Y.; Li, L.; Zhang, C.; Casillas, G.; Sun, Z.; Yan, Z.; Ruan, G.; Peng, Z.; Raji, A.-R. O.; Kittrell, C.; Hauge, R. H.; Tour, J. M. A Seamless Three-Dimensional Carbon Nanotube Graphene Hybrid Material. *Nat. Commun.* **2012,** *3,* 1225.
11. Kim, S. H.; Song, W.; Jung, M. W., Kang, M.-A.; Kim, K.; Chang, S.-J.; Lee, S. S.; Lim, J.; Hwang, J.; Myung, S.; An, K.-S. Carbon Nanotube and Graphene Hybrid Thin Film for Transparent Electrodes and Field Effect Transistors. *Adv. Mater.* 2014, *26* (25), 4247–4252.
12. Hsieh, C.-T.; Liu, Y.-Y.; Tzou, D.-Y.; Chen, W.-Y. Atomic Layer Deposition of Platinum Nanocatalysts onto Three-Dimensional Carbon Nanotube/Graphene Hybrid. *J. Phys. Chem. C* **2012,** *116*, 26735–26743.
13. Chen, S.; Yeoh, W.; Liu, Q.; Wang, G. Chemical-Free Synthesis of Graphene–Carbon Nanotube Hybrid Materials for Reversible Lithium Storage In Lithium-Ion batteries. *Carbon* **2012,** *50* (12), 4557–4565.
14. Dichiara, A.; Yuan, J.-K.; Yao, S.-H.; Sylvestre, A.; Bai, J. Chemical Vapor Deposition Synthesis of Carbon Nanotube-Graphene Nanosheet Hybrids and Their Application in Polymer Composites. *J. Nanosci. Nanotechnol.* **2012,** *12*, 6935–6940.
15. Mani, V.; Chen, S.-M.;Lou, B.-S. Three Dimensional Graphene Oxide-Carbon Nanotubes and Graphene-Carbon Nanotubes Hybrids. *Int. J. Electrochem. Sci.* **2013,** *8*, 11641–11660.
16. Wu, Y.; Zhang, T.; Zhang, F.; Wang, Y.; Ma, Y.; Huang, Y.; Liu, Y.; Chen, Y. In Situ Synthesis of Graphene/Single-Walled Carbon Nanotube Hybrid Material by Arc-Discharge and Its Application in Supercapacitors. *Nano Energy* **2012,** *1*, 820–827.
17. Sui, Z.; Meng, Q.; Zhang, X.; Mab, R.; Cao, B. Green Synthesis of Carbon Nanotube–Graphene Hybrid Aerogels and Their Use as Versatile Agents for Water Purification. *J. Mater. Chem.* **2012**, *22*, 8767.
18. Tung, V. C.; Chen, L. M.; Allen, M. J.; Wassei, J. K.; Nelson, K.; Kaner, R. B.; Yang, Y. Low-Temperature Solution Processing of Graphene–Carbon Nanotube Hybrid Materials for High-Performance Transparent Conductors. *Nano Lett.* **2009,** *9* (5), 1949–1955.

19. Cai, D.; Song, M.; Xu, C. Highly Conductive Carbon-Nanotube/Graphite-Oxide Hybrid Films. *Adv. Mater.* **2008,** *20* (9):1706–1709.
20. Huang, J. H.; Fang, J. H.; Liu, C. C.; Chu, C. W. Effective Work Function Modulation of Graphene/Carbon Nanotube Composite Films as Transparent Cathodes for Organic Optoelectronics. *ACS Nano* **2011,** *5*, 6262.
21. Song, M.; Xu, P.; Song, Y.; Wang, X.; Li, Z.; Shang, X.; Wu, H.; Zhao, P.; Wang, M. Water-Processed Carbon Nanotube/Graphene Hybrids with Enhanced Field Emission Properties. *AIP Adv.* **2015,** *5*, 097130.
22. King, P. J.; Khan, U.; Lotya, M.; De, S.; Coleman, J. N. Improvement of Transparent Conducting Nanotube Films by Addition of Small Quantities of Graphene. *ACS Nano* **2010,** *4*, 4238–4246.
23. Kim, Y. K.; Na, H. K.; Kwack, S. J.; Ryoo, S. R.; Lee, Y. M.; Hong, S. H.; et al. Synergistic Effect of Graphene Oxide/MWCNT Films in Laser Desorption/Ionization Mass Spectrometry of Small Molecules and Tissue Imaging. *ACS Nano* **2011,** *5*, 4550–4561.
24. Hong, T. K.; Lee, D. W.; Choi, H. J.; Shin, H. S.; Kim, B. S. Transparent, Flexible Conducting Hybrid Multilayer Thin Films of Multiwalled Carbon Nanotubes with Graphene Nanosheets. *ACS Nano* **2010,** *4* (7), 3861–3868.
25. Yu, D.; Dai, L. Self-Assembled Graphene/Carbon Nanotube Hybrid Films for Supercapacitors. *J. Phys. Chem. Lett.* **2010,** *1* (2), 467–470.
26. Diamanti, E. K.; Falzon, B. G.; Hawkins, S. C. In *Comparative Study of Electrical Properties of Different Carbon Nanotube Buckypaper for Lightning Strike Protection Applications*, 21st International Conference on Composite Materials Xi'an, 20-25th August 2017.
27. Fan, W.; Miao, Y.-E.; Zhang, L.; Huang, Y.; Liu, T. Porous Graphene–Carbon Nanotube Hybrid Paper as a Flexible Nano-Scaffold for Polyaniline Immobilization and Application in All-Solid-State Supercapacitors. *RSC Adv.* **2015,** *5*, 31064.
28. Sasikaladevi, S.; Aravind, J.; Eswaraiah, V.; Ramaprabhu, S. Facile Synthesis of One Dimensional Graphene Wrapped Carbon Nanotube Composites by Chemical Vapour Deposition. *J. Mater. Chem.* **2011,** *21*, 15179.
29. Kaushik, V.; Shukla, A. K.; Vankar, V. D. Microwave Plasma CVD-Grown Graphene—CNT Hybrids for Enhanced Electron Field Emission Applications. *Appl. Phys. A* **2014,** *117*, 2197–2205.
30. Zhou, Y.; Qian, W.; Huang, W.; Liu, B.; Lin, H.; Dong, C. Carbon Nanotube-Graphene Hybrid Electrodes with Enhanced Thermo-Electrochemical Cell Properties. *Nanomaterials* **2019,** *9*, 1450.
31. Lin, C.-C.; Lin, Y.-W. Synthesis of Carbon Nanotube/Graphene Composites by One-Step Chemical Vapor Deposition for Electrodes of Electrochemical Capacitors. *J. Nanomater.* **2015,** *2015*, 741928.
32. Koh, A. T. T.; Chen, T.; Pan, L.; Sun, Z.; Chua, D. H. C. Effective Hybrid Graphene/Carbon Nanotubes Field Emitters by Electrophoretic Deposition. *J. Appl. Phys.* **2013,** *113*, 174909.
33. Muangrat, W.; Wongwiriyapan, W.; Morimoto, S.; Hashimoto, Y. Graphene Nanosheet-Grafted Double-Walled Carbon Nanotube Hybrid Nanostructures by Two-Step Chemical Vapor Deposition and their Application for Ethanol Detection. *Sci. Rep.* **2019,** *9*, 7871.

34. Jyothirmayee Aravind, S. S.; Imran Jafri, R.; Rajalakshmib, N.; Ramaprabhu, S.; Solar Exfoliated Graphene–Carbon Nanotube Hybrid Nano Composites as Efficient Catalyst Supports for Proton Exchange Membrane Fuel Cells. *J. Mater. Chem.* **2011,** *21*, 18199.
35. Du, H. Y.; Wang, C. H.; Hsu, H. C.; Chang, S. T.; Huang, H. C.; Chen, L. C.; Chen, K. H. Graphene Nanosheet—CNT Hybrid Nanostructure Electrode for a Proton Exchange Membrane Fuel Cell. *Int. J. Hydrog. Energy* **2012,** *37* (24), 18989–18995.
36. Zou, L.; Qiao, Y.; Wu, X. S.; Li, C. M. Tailoring Hierarchically Porous Graphene Architecture by Carbon Nanotube to Accelerate Extracellular Electron Transfer of Anodic Biofilm in Microbial Fuel Cells. *J. Power Sources* **2016,** *328*, 143–150.
37. Wu, X.; Qiao, Y.; Shi, Z.; Tang, W.; Li, C. M. Hierarchically Porous N-Doped Carbon Nanotubes/Reduced Graphene Oxide Composite for Promoting Flavin-Based Interfacial Electron Transfer in Microbial Fuel Cells. *ACS Appl. Mater. Interfaces* **2018,** *10* (14), 11671–11677.
38. Prasad, K. P.; Chen, Y.; Chen, P. Three-Dimensional Graphene-Carbon Nanotube Hybrid for High-Performance Enzymatic Biofuel Cells. *ACS Appl. Mater. Interfaces* **2014,** *6* (5), 3387–3393.
39. Aval, L. F.; Ghoranneviss, M.; Pour, G. B. High-Performance Supercapacitors Based on the Carbon Nanotubes, Graphene and Graphite Nanoparticles Electrodes. *Heliyon* **2018,** *4* (11), e00862.
40. Ansaldo, A.; Bondavalli, P.; Bellani, S.; Del Rio Castillo, A. E.; Prato, M.; Pellegrini, V.; Pognon, G.; Bonaccorso, F. High-Power Graphene–Carbon Nanotube Hybrid Supercapacitors. *Chem. Nano Mat.* **2017,** *3* (6), 436–446.
41. Aphale, A.; Maisuria, K.; Mahapatra, M. K.; Santiago, A.; Singh, P.; Patra, P. Hybrid Electrodes by In-Situ Integration of Graphene and Carbon-Nanotubes in Polypyrrole for Supercapacitors. *Sci. Rep.* **2015,** *5* (1), 1–8.
42. Jung, N.; Kwon, S.; Lee, D.; Yoon, D. M.; Park, Y. M.; Benayad, A.; Choi, J. Y.; Park, J. S. Synthesis of Chemically Bonded Graphene/Carbon Nanotube Composites and their Application in Large Volumetric Capacitance Supercapacitors. *Adv. Mater.* **2013,** *25* (47), 6854–6858.
43. Jiang, L.; Sheng, L.; Long, C.; Fan, Z. Densely Packed Graphene Nanomesh-Carbon Nanotube Hybrid Film for Ultra-High Volumetric Performance Supercapacitors. *Nano Energy* **2015,** *11*, 471–80.
44. Fan, Z.; Yan, J.; Zhi, L.; Zhang, Q.; Wei, T.; Feng, J.; Zhang, M.; Qian, W.; Wei, F. A Three-Dimensional Carbon Nanotube/Graphene Sandwich and Its Application as Electrode in Supercapacitors. *Adv. Mater.* **2010,** *22* (33), 3723–3728.
45. Zheng, Q.; Cai, Z.; Ma, Z.; Gong, S. Cellulose Nanofibril/Reduced Graphene Oxide/Carbon Nanotube Hybrid Aerogels for Highly Flexible and All-Solid-State Supercapacitors. *ACS Appl. Mater. Interfaces* **2015,** *7* (5), 3263–3271.
46. Wang, W.; Guo, S.; Lee, I.; Ahmed, K.; Zhong, J.; Favors, Z.; Zaera, F.; Ozkan, M.; Ozkan, C. S. Hydrous Ruthenium Oxide Nanoparticles Anchored to Graphene and Carbon Nanotube Hybrid Foam for Supercapacitors. *Sci. Rep.* **2014,** *4* (1), 1–9.
47. Sun, G.; Zhang, X.; Lin, R.; Yang, J.; Zhang, H.; Chen, P. Hybrid Fibers Made of Molybdenum Disulfide, Reduced Graphene Oxide, and Multi-Walled Carbon Nanotubes for Solid-State, Flexible, Asymmetric Supercapacitors. *Angew. Chem. Int. Ed.* **2015,** *127* (15), 4734–4739.

48. Liu, Y.; Liu, Y.; Qin, S.; Xu, Y.; Zhang, R.; Wang, F. Graphene-Carbon Nanotube Hybrid Films for High Performance Flexible Photodetectors. *Nano Res.* 2017, *10*, 1880–1887.
49. Kuhn, L.; Gorji, N. E. Review on the Graphene/Nanotube Application in Thin Film Solar Cells. *Mater. Lett.* **2016,** *171*, 323–326.
50. Kavadiya, S.; Niedzwiedzki, D. M.; Huang, S.; Biswas, P. Electrospray-Assisted Fabrication of Moisture-Resistant and Highly Stable Perovskite Solar Cells at Ambient Conditions. *Adv. Energy Mater.* **2017,** *7*, 1700210.
51. Li, X.; Tong, T.; Wu, Q.; Guo, S.; Song, Q.; Han, J.; Huang, Z. Unique Seamlessly Bonded CNT@Graphene Hybrid Nanostructure Introduced in an Interlayer for Efficient and Stable Perovskite Solar Cells. *Adv. Funct. Mater.* **2018,** *28* (32), 1800475.
52. Salim, T.; Lee, H.-W.; Wong, L. H.; Oh, J. H.; Bao, Z.; Lam, Y. M. Semiconducting Carbon Nanotubes for Improved Efficiency and Thermal Stability of Polymer—Fullerene Solar Cells. *Adv. Funct. Mater.* **2016,** *26*, 51.
53. Yang, Z.; Liu, M.; Zhang, C.; Tjiu, W. W.; Liu, T.; Peng, H. Carbon Nanotubes Bridged with Graphene Nanoribbons and Their Use in High-Efficiency Dye-Sensitized Solar Cells. *Angew. Chem. Int. Ed.* **2013,** *52*, 3996–3999.
54. Lee, K.-T.; Park, D. H.; Baac, H. W.; Han, S. Graphene- and Carbon-Nanotube-Based Transparent Electrodes for Semitransparent Solar Cells. *Materials* **2018,** *11*, 1503.

CHAPTER 2

GRAPHENE-BASED MATERIALS FOR ENERGY STORAGE AND CONVERSION APPLICATION

ASHUTOSH MAJHI, LIPSA SHUBHADARSHINEE, PATITAPABAN MOHANTY, BIGYAN RANJAN JALI, PRIYARANJAN MOHAPATRA, and ARUNA KUMAR BARICK*

Department of Chemistry, Veer Surendra Sai University of Technology, Siddhi Vihar, Burla, Sambalpur 768018, Odisha, India

Corresponding author. E-mail: akbarick_chem@vssut.ac.in

ABSTRACT

Ongoing rapid depletion of global energy attributed to the improvement of immaculate and renewable energy sources become indispensable. To face the challenge of this energy caused by those newly introduced molecules, nanotechnology comes up with new frontiers in engineering and material science especially carbon nanomaterials as it has capability of efficient energy storage and conversion. Graphene as a novel 2D nanomaterial has great potential with many remarkable applications and also graphene-based assemblies are put forth in the area of research owing to its inherent physicochemical properties as well as an optimistic material. It has several applications like conversion of energy and storage of energy. This book chapter thoroughly goes through the development and progress of graphene and graphene-based material in the field of research for the improvement of energy storage equipment, such as supercapacitors, lithium-ion batteries, and energy-conversion devices, including solar cells,

Nanostructured Carbon for Energy Generation, Storage, and Conversion. V. I. Kodolov, DSc, Omari Mukbaniani, DSc, Ann Rose Abraham, PhD, and A. K. Haghi, PhD (Eds.)

fuel cells. Various structures and modified mechanisms are expected just to increase the potentiality of graphene-based material, that is, to increase their capacity of storing energy and conversion, respectively. The advantages, challenges, and limitation of graphene-based materials, when they are being utilized in different energy-related aspects in terms of storage and conversion, have also been focused in this book chapter.

2.1 INTRODUCTION

Currently, the world faces some leading challenges regarding consumption, storage, and production of viable energy. Except assembling the sources of workable and renewable energy, we also have to be aware about the efficient storage of energy, so that we can supply it at the time of demand mainly in various applications, such as portable electronic devices and transportation systems. There are three key ways by which energy can be stored which include chemically, electrically, and electrochemically. There are huge numbers of materials available, which can be used for energy storage.[1] To deliver the extensive use of renewable sources, there must be requirement of well-organized technology in the field of conversion and energy storage processes.[2]

Graphene, after receiving Novel prize in the field of physics on 2010 for "ground-breaking experiments regarding the two-dimensional (2D) material graphene" has received an explosion of interest from researchers. After that, several patents along with papers based on the graphene and synthesis of graphene-based material and their applications are still increasing day by day.[3]

Graphene is an allotrope of carbon and has been arranged in a hexagonal manner by bonding with sp^2 hybridized carbon, and it has an indefinitely extended two-dimensional (2D) monolayer of sp^2 carbon sheets. Graphene is transparent to electrons, and it is practically impermeable to all molecules at room temperature. Figure 2.1 shows the crystal lattice structure of graphene.[4]

Though graphene is a constituent of three-dimensional (3D) materials, it is assumed not to be remaining in free state and initially has been considered to be an "academic" material and that of corresponding to the curve structures like nanotubes and fullerenes, it was considered to be unstable.

In other words, graphene has a flat monolayer that was densely packed with carbon moieties and being considered a basic constituent for all other more dimensional graphitic materials. Graphene is enfolded into a zero-dimensional (0D) fullerenes (wrapped graphene), rolled into a one-dimensional (1D) nanotubes (rolled graphene) and stacked into a three-dimensional (3D) graphite (stacked graphene) as shown in Figure 2.2.

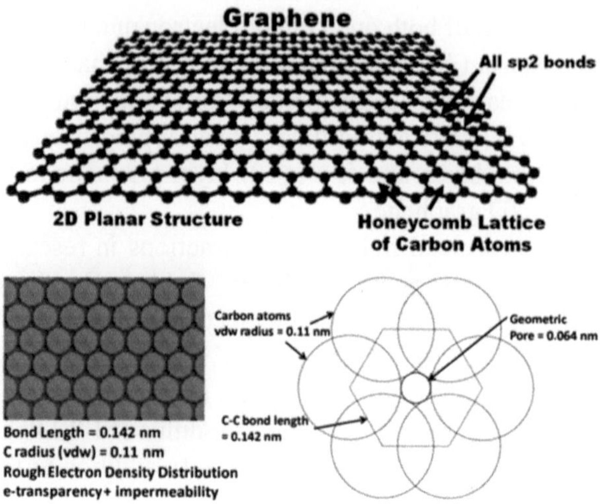

FIGURE 2.1 Schematic representation of the crystal lattice structure of graphene.[4]
Source: Reprinted with permission from Ref. [4]. © 2013 Elsevier.

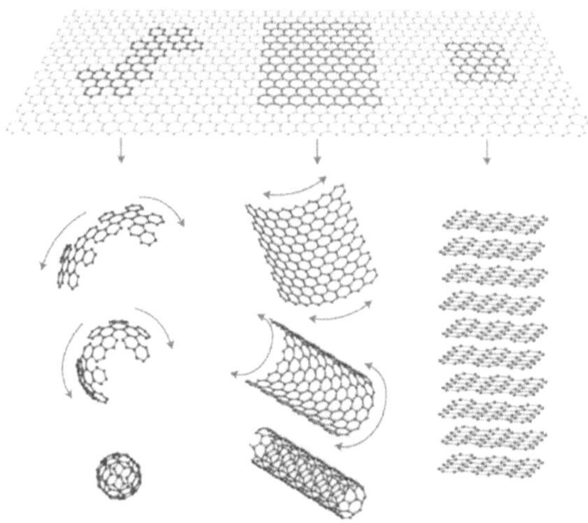

FIGURE 2.2 Schematic representation of the mother of all graphitic forms.[5]
Source: Reprinted with permission from Ref. [5]. © 2007 Springer.

The electronic structure quickly changes with the number of layers, which in the case of graphite, is approaching the 3D limits with 10 carbon layers. Furthermore, simply graphene, that has been taken into a good assessment, its two layer shows a plain electronic spectrum, that is, it is a zero-band gap semiconductor and similarly acts as a zero-overlap metalloids consisting of both one kind of electron and hole. For more than bilayers, the spectrum become gradually complex, that is, appearance of numerous charge carriers occurs thereby the conduction bands (CB) and valence bands (VB) start extraordinary overlapping. Hence, this admits single layer, double layer, and few (≤10)-layer graphene to be prominent as three types of 2D graphene.[5]

Graphene has received tremendous attractions in research and widespread applications in multidisciplinary fields.[2] Graphene possesses unique properties, which includes high electrical conductivity (10^8 Sm^{-1}), great thermal conductivity (~5000 Wm^{-1} K^{-1}), excellent optical transparency (optical density (OD) of 2.3 %), good mechanical properties (tensile strength at break of 42 Nm^{-1} and Young's modulus of 1.0 TPa), large aspect ratio, essential elasticity, great surface area.[6] In addition to this, graphene also has some extraordinary properties, such as ambipolar electric field effect, massless electron, very large transport number of charge carriers (up to 10^5 $cm^2V^{-1}s^{-1}$), display of the quantum hall effect (QHE), and propagation of electronic waves inside one atomic layer. Graphene also exhibits excellent intrinsic strength (~130 GPa) and Young's modulus, which make it the strongest material ever investigated. Graphene possesses very high surface area (~2600 m^2g^{-1}) that is appreciably larger than that of the surface area of graphite (~10 m^2g^{-1}) and carbon nanotubes (CNTs) (~1300 m^2g^{-1}). These outstanding properties of graphene make it a suitable material for different uses, such as paper-like materials, polymer nanocomposite materials, field effect transistors, photoelectronics, sensors and probes, electromechanical systems, hydrogen storage, and electrochemical energy systems.[7]

Recently, graphene material has come up with one of its enormous properties, that is, as a substituent for energy storage. The great surface area of graphene is highly favorable to apply in energy storage. Graphene is an electrical conducting material and can be easily functionalized by means of different chemical modification methods. By comparing the chemical derivatives and structural aspect of graphene, researcher represents "graphenes" as a family of graphene-related materials. Graphene resembles an extent range of polyaromatic compound as it has only limited to exit in-plane, so it was called as nanoribbon of graphene having one, two, or more than two layers. Those multilayer nanoribbons of graphene are considered as assembled graphene nanofibers or nanofibers of graphite. The oxides of graphene are exclusively significant derivatives of graphene, prepared chemically, that is, it contains a single-layered graphite oxide. Graphene oxide (GO) is synthesized through simple chemical processes, that is, after graphite is oxidized to oxides of graphite it simply goes through sonication and exfoliation (stirring for a long period) under thermal condition respectively.

Generally, there are three strategies employed to design the morphology and performance of the graphene, which includes modification, functionalization, and self-assembly of the graphene nanosheets (GNSs).

2.1.1 STRUCTURAL MODIFICATION OF GRAPHENE

Graphene shows notable electronic properties having a lateral dimension varying from 100 nm to 10 μm. However, graphene is a metalloid that may not be generally utilized for fabrication of nanoscopic semiconductor devices (for example, field effect transistors) because of the absence of band gap. Additionally, the structural modification of the graphene surface encourages both *quantum confinement* and *edge effects*, which manipulate the transportation of charge carriers and optical properties. Microscopic structural modification of graphene, including graphene quantum dot (GQD), graphene nanoribbon (GNR), and graphene nanomesh (GNM) is shown in Figure 2.3A.

2.1.2 FUNCTIONALIZATION OF GRAPHENE

Figure 2.3B represents the functionalization of graphene, which is an essential route to modify the performance of the graphene. The functionalization process is generally classified into the following two types. Firstly, by purposely incorporating heteroatoms (for example, boron, nitrogen, phosphorus, sulfur, fluorine, chlorine) into the hexagonal crystal *lattice of* graphene nanolayers that results deformations of both structure and electronic distributions. Consequently, the intrinsic band structure is modified and the optical, magnetic, and electrical properties of the graphene is successfully regulated. Secondly, fabrication of graphene-based hybrid materials by using metals, metal oxides, metal sulfides, polymers, carbon polyhedrons, carbon nitrides.

2.1.3 SELF-ASSEMBLY OF GRAPHENE

Development of macroscale graphene-based architectural designs by taking GNSs as primary building block is very effective approach pertaining to the scalable applications of graphene. On the other hand, hydrophilic GO can be derived from graphene through self-assemble process. Furthermore, it can be reduced into the different nanostructures because of the presence of large number of different oxygen-containing functional groups, such as hydroxyl, epoxy, and carboxyl both at the basal plane and at sheet edges. Figure 2.3C represents the graphene fibers/microtubules, graphene films, and graphene foam derived from the self-assembly of GNSs.

Thus, the functionalization and self-assembly of graphene offer a suitable practical approach to synthesize numerous graphene-based nanomaterials on an industrial scale that open an opportunity for successful applications of graphenes.[8]

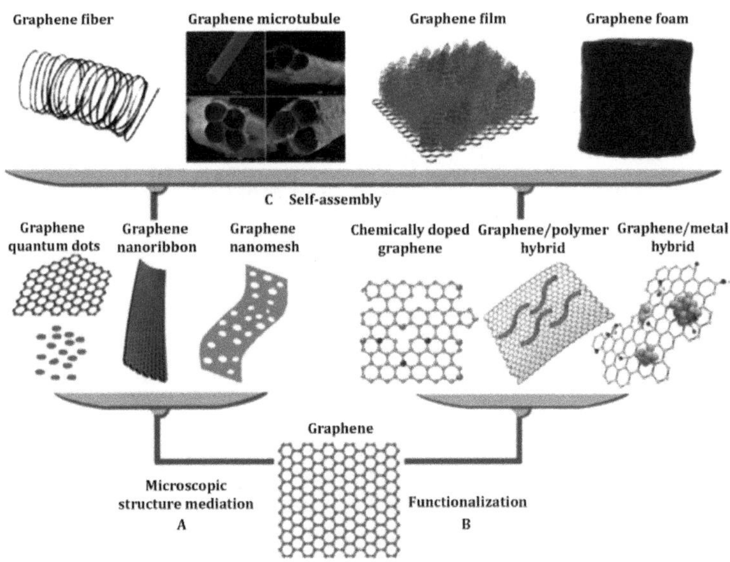

FIGURE 2.3 Schematic representation of the progress in tailoring various types of graphene.[8]
Source: Reprinted with permission from Ref. [8]. © 2018 Elsevier.

Graphene is attracting many attentions due to its excellent material properties and having large possibility for several end uses. Predominantly, applications of graphene are cost-effective and competent candidate for energy conversion and storage mechanisms, such as supercapacitors, *lithium-ion battery,* solar cells, fuel cells. However, graphene needs to be produced in huge amounts for practical applications. The accessibility and processing of graphene has rate-controlling steps for its exact assessment.[9] The spatial arrangement and number of graphene nanolayers are accountable for the electronic behavior of the graphene. The electronic structure of few-layer graphene is unlike bulk graphite. The chemical and physical characteristics of graphene molecules can be affected by that bulkiness and that of presence of multiple layers as well as by ordering of those intermediating layers. Graphene because of its extraordinary properties, that is, its 2D nanostructures can be converted into the 3D network structures and all other properties have received a great attention.[10] To develop extremely efficient energy storage (e.g., supercapacitor and Li-ion battery) and energy conversion (e.g., solar cell and fuel cell) systems, reasonable efforts have been made. The implementation of these systems directly governed by the efficiency of electrode materials and their properties.[11]

2.2 LITERATURE SURVEY

In 1859, British chemist Brodie obtained "carbonic acid" (Fig. 2.4a) by exposing graphite to strong acids. Brodie thought that he discovered "graphon," which is a type of carbon having mol. wt. 33. Nowadays, it is known that he detected a suspensions of minute crystals of GO, which is defined as a graphene nanosheet compactly enclosed with hydroxyl and epoxide groups.[12,13] In 1948, Ruess and Vogt, after putting a drop of GO suspensions on a transmission electron microscopy (TEM) grid, have detected wrinkled pieces of nanometer thickness.[12,14] In 1962, Boehm et al. searched the tiniest plausible fragment of reduced GO (RGO) and recognized them as single nanolayers (Fig. 2.4b).[12,15] The term "graphene" was coined by Boehm et al. in 1986 which is invented by the blending of the word "graphite" and the suffix "ene" that represents to polycyclic aromatic hydrocarbons (PAHs). Forty years after the 1962 paper, graphene monolayers were clearly identified in TEM by counting the number of folding lines.[12,16]

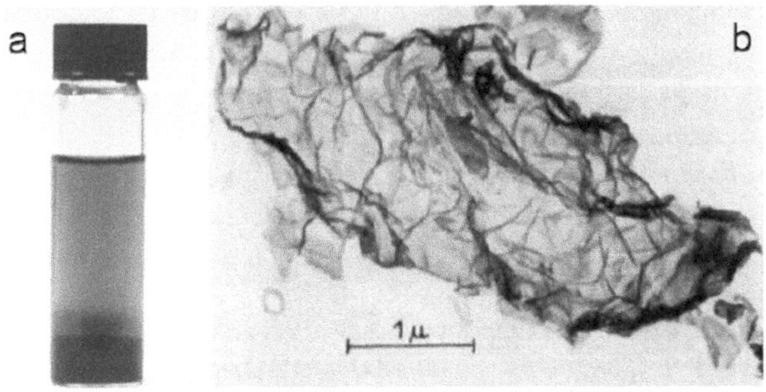

FIGURE 2.4 (**a**) Photograph of GO slurry and (**b**) TEM image of graphene.[12]
Source: Reprinted with permission from Ref. [12]. © 1970 IOP Publishing.

In 2004, at Manchester University, Geim and colleagues first isolated single-layer samples from graphite (Fig. 2.5).[17,18]

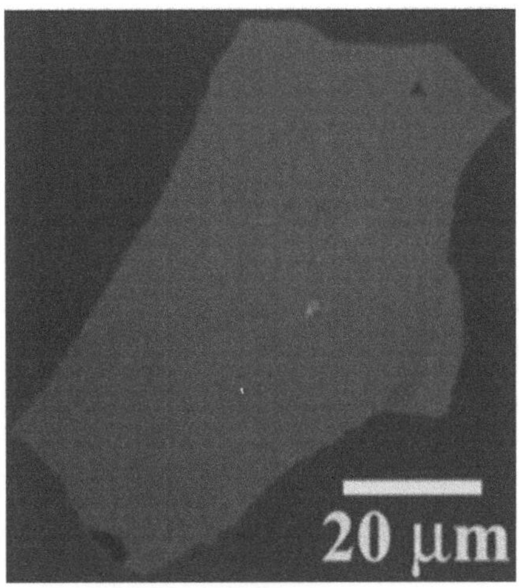

FIGURE 2.5 TEM image of single-layer graphene.[18]
Source: Reprinted with permission from Ref. [18]. © 2010 American Chemical Society.

In graphene, the long-range π-conjugation yields unique mechanical, thermal, electrical properties that become interesting in studies and lately received great attention for experiment research. The curiosity on graphene in many researchers is not due to its synthesis and morphology but because of its electronic properties.[12,18]

2.3 SYNTHESIS OF GRAPHENE

Graphene is synthesized mainly by two approaches, such as top-down and bottom-up methods. Top-down method accounts for the separation of stacked graphite layer to produce single graphene sheets (GS) and bottom-up method includes the preparation of graphene from different carbon precursors. Recent advancement on the synthesis of graphene and graphene-based materials has further classified synthesis into several methods. The top-down method involves liquid-phase exfoliation, electrochemical exfoliation, and chemical reduction of GO but the bottom-up method includes epitaxial growth technique, chemical vapor deposition (CVD), and chemical synthesis from aromatic sources. The schematic representation for the preparation of graphene through top-down and bottom-up methods is shown in Figure 2.6.[3]

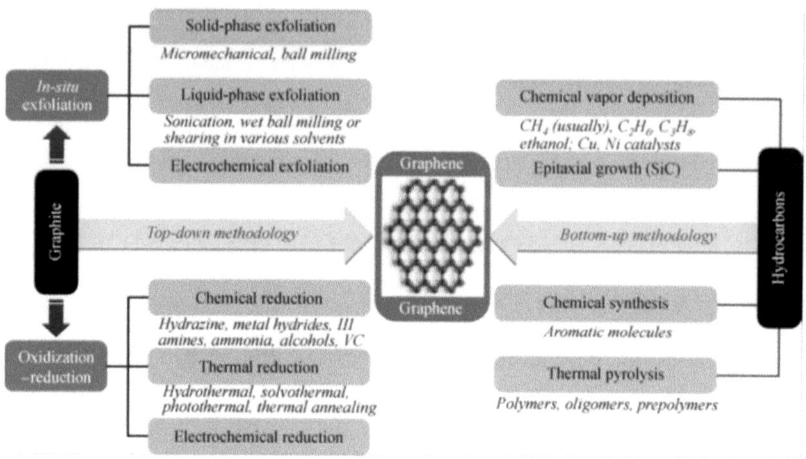

FIGURE 2.6 Schematic representation for the synthesis of graphene.[19]

Source: Reprinted with permission from Ref. [19]. © 2016 Elsevier.

The methods implemented for graphene production perform a vital factor in governing the nature and performance of the final product. Every single method is estimated in terms of quality, cost-effectiveness, large-scale processability, purity, and yield of the production of graphene. In agreement with these characteristics, the most common graphene production methods are shown in Figure 2.7a.[20] Various methods implemented for large-scale production of graphene that allows a widespread option based on the size, quality and cost for any specific functions are represented in Figure 2.7b.[21]

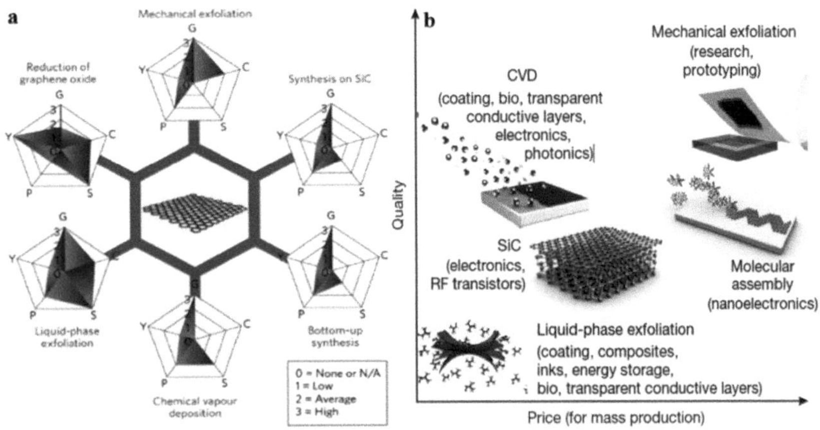

FIGURE 2.7 Schematic representation of (a) most common and (b) industrial production methods for graphene.[20,21]

Source: Reprinted with permission from a) Ref. [20]. © 2014 Springer; b) Ref. [21]. © 2012 Springer.

2.4 GRAPHENE-BASED MATERIALS FOR ENERGY STORAGE

2.4.1 *SUPERCAPACITOR*

Supercapacitors are generally considered to be an energy storage machines which have capability of energy storage and a tendency of liberating that energy within a limited time period with a high accuracy of power, current density, and long cycle stability. Supercapacitors are also referred as ultracapacitors or electrochemical capacitors. They have higher storage capacities than that of regular dielectric capacitors but comparatively less than batteries. The principle of energy storage in a supercapacitor operates by two mechanisms. The first one is by doubly layered electrical capacitance which was being operated by applying ions produced from continuous absorption and desorption processes on electrolytic interface forming a double-layer structure. The second type is pseudocapacitance, which utilizes transition metal oxides or conducting polymers as electrode materials where the charge storage depends on rapid Faradic redox reaction. It is noteworthy that most of the supercapacitors work by combination of both mechanisms. For electric double-layer capacitor (EDLC), the capacitance is directly proportional to surface area, dielectric constant and inversely proportional to double-layer thickness. The amalgamation of large surface area and little charge separation is essential for superior capacitance. On the other hand, for pseudocapacitors, the capacitance concerns voltage supported reversible Faradic reactions among electrodes with electrolytes through adsorption and desorption of charge carriers on the surface or redox reactions with electrolytes or doping and undoping of electrodes. The amount of energy accumulated in a supercapacitor (E) is measured using the formula $E = 1/2CV^2$, where C is capacitance and V is working voltage.[1,7,22] In EDLC, the active materials should not only be stable, but also should have accessible large surface area. Additionally, there should not be Faradic reaction at EDLC electrode. Thus, the charge storage on surface grants rapid energy uptake and release thereby leading to superior functioning. Alternatively, pseudocapacitors undergo reversible Faradic reaction and the stability for charge/discharge cycling is comparatively low. However, their energy densities are more in comparison with EDLC. Moreover, the response time in pseudocapacitor is longer than EDLC, because the time taken to move electrons during redox reaction is more.[23]

The schematic representation of charge storage in a supercapacitor by EDLC, pseudocapacitance, and hybrid capacitance mechanism is shown in Figure 2.8(a–c) and 2.8(d) represents typical charge–discharge voltammetry characteristics of a supercapacitor.

Nanostructures provide large surface area and increase availability of dynamic materials for the process of electrolysis by which there was formation of multiple layer and also involvement of pseudocapacitor for reaction to take place.[22] Graphene is regarded as a competitive candidate for supercapacitor application as a result of its excellent material properties, which further increases the studies and research on graphene-based supercapacitor. The effective surface area of graphene relies on the layers present in it. With less agglomeration, single- or few-layer graphene can be presumed to show a large effective surface area thereby increasing the supercapacitor properties.[9] Based on electrode materials, graphene-based supercapacitors mostly consist of neat graphene-based EDLC, graphene and metal oxide or chalcogenides nanocomposites and graphene/conducting polymer nanocomposite-based pseudocapacitor.[11] In addition to this, the surface area of GS cannot change with porous size distribution and the GS make available of both surfaces to electrolytes. A single-layer graphene, which possesses little cluster, is anticipated to show large surface area and therefore results in higher specific capacitance of supercapacitors. However, stacked GS form an aggregated structure that may contain pores. Functioning of graphene-based supercapacitor can be improved via preparation of graphene/metal oxide composites, modification of graphene surface or by forming hybrid structure with electrically conductive polymers.[7]

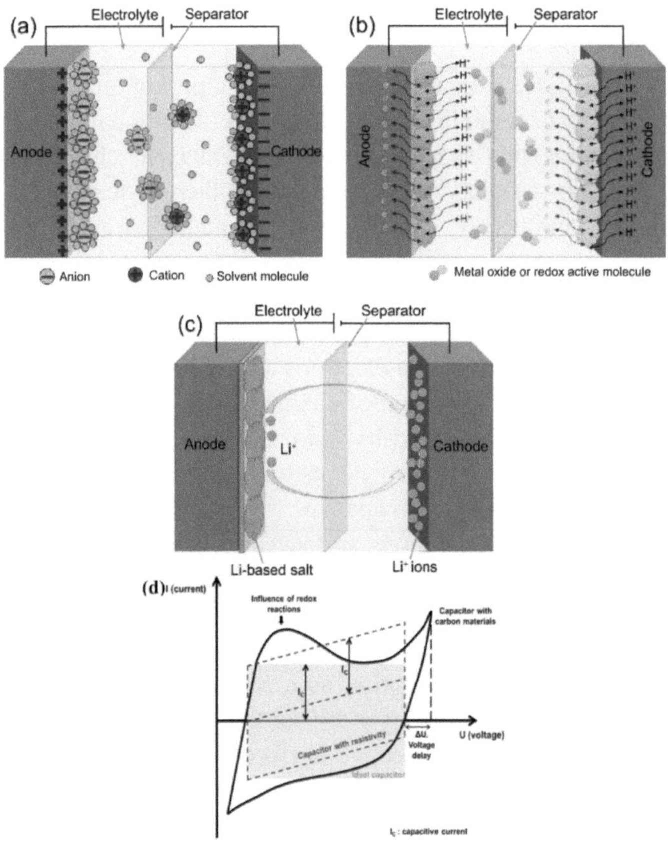

FIGURE 2.8 Schematic representation of charge storage in (a) electrical double-layer capacitor, (b) pseudocapacitor, (c) hybrid supercapacitor,[24] and (d) typical charge–discharge voltammetry characteristics of a supercapacitor.[23]

Source: Reprinted with permission from (a-c) Ref. [24]. © 2017 Oxford University Press; (d) Ref. [23]. © 2012 Elsevier.

EDLC is directly proportional to the surface area and though graphene has maximum area that of nearly 2630 m^2 g^{-1}, but it is involved in supercapacitors as a novel medium. Stoller et al. first reported this research. It was known that chemically modified graphene (CMG) shows high capacitance of 99 Fg^{-1} and 135 Fg^{-1} for corresponding organic-based electrolytes and aqueous electrolyte respectively, and this is shown in Figure 2.9. SEM image of CMG particle surface is shown in Figure 2.9a, TEM image of individual GS extending from CMG particle is shown in Figure 2.9b. Furthermore, Figure 2.9c represents low, high (inset) resolution of SEM image of CMG electrode surface, and Figure 2.9d represents diagram for fabrication of test cell.[25]

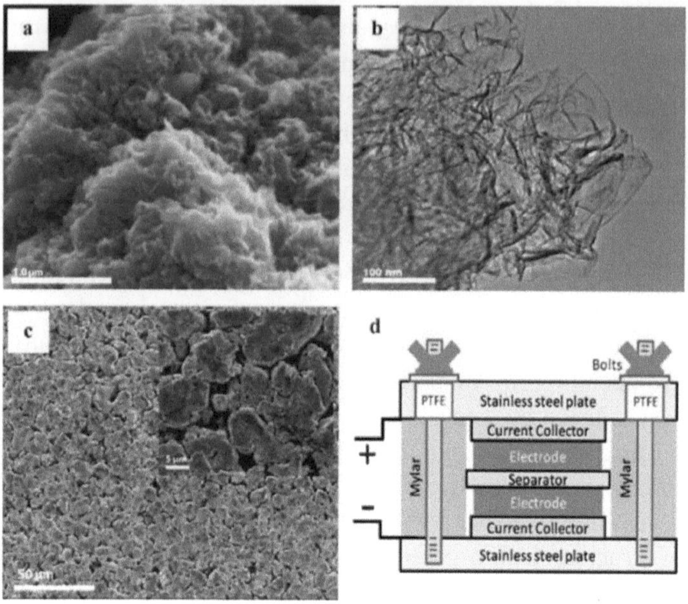

FIGURE 2.9 (a) SEM microphotograph of CMG, (b) TEM microphotograph of GS, (c) TEM microphotograph of CMG electrode surface, and (d) schematic representation of test cell.[25]

Source: Reprinted with permission from Ref. [25]. © 2008 American Chemical Society.

Wang et al. further reported about those graphene materials, that is, their capacitance increases by a factor of 205 Fg^{-1}.[26] He proposed that there was increase in the value of Capacitance by a factor of 255 Fg^{-1}, that is, from 14 Fg^{-1} to 269 Fg^{-1} was only due to the spacing of layers within graphene molecule by nanocrystals of Pt, having diameter of 4 nm. To increase the area, which is available electrochemically and correspondingly to minimize the clustering of those graphene material, GS can be used, as they have high pseudocapacitance. The capacitances shown by GO and bulk MnO_2 are 10.9 Fg^{-1} and 6.8 Fg^{-1}, respectively, but MnO_2 nanocrystals, which were being synthesized on the sheets of GO as a result a long-range capacitance of 197 Fg^{-1} were observed.[1] Du et al. prepared two types of GS derivative by the process of thermal exfoliation of graphitic oxide at minimal temperature in the presence of normal atmosphere and N_2 in high temperature. Specific capacitance value of functionalized GS derived from low temperature is 230 Fg^{-1}, which is almost twice as much as the value of high temperature, type that is 100 Fg^{-1}.[9,27] Author Lv also prepared graphene molecule by exfoliating the graphite oxide at temperature nearly 200°C (low range) at vacuum. The newly formed graphene has a long-range capacitance of 264 Fg^{-1}.[28] This phenomenon is attributed the presence of various activated groups, such as epoxy, carboxyl, and hydroxyl groups upon the reactive surface, which results in fast redox reaction happening over the electrode surface that offers pseudocapacitance, and thus increases the charge storage. A three-dimensional CGS along with CNT base was being developed in-between the layers of graphene and also fabricated and being operated as a prime electrode of supercapacitors. When the ions of electrolyte, that is, electrons are being transported rapidly within the sandwich-shaped structure, it causes complete operation of double-layer capacitance from graphene along with pseudocapacitance by catalyst, thus a long range of capacitance was being recorded that is 385 Fg^{-1} in 6 molar potassium hydroxide solution.[9] To stop the rearrangement of newly derived GS through chemical process, author Li introduced a multiple layered film of graphene having widely open arrangement of structure with a potent spacer like water molecule so that it can show easy availability of electrolytic solution on the plane of each sheets. Those graphene films have certain capacitance range of nearly 215 Fg^{-1} in electrolyte solution having aqueous phase.[29] Furthermore, the supercapacitor having ionic liquid-exchanged film has a capacitance of 273.1 Fg^{-1}. More interestingly, a kind of, one-dimensional nanomaterial with core construction of graphene has been prepared. This carbonaceous graphene has high porosity and being prepared under high range of temperature through the process of pyrolysis with the percussing agent like carbon abundant group polyphenylene. At 800 °C, when a graphene-based molecule was being produced with a specified surface area (SSA) of maximum 1140 m^2g^{-1} was taken as electrode in EDLC, it yields a higher specific capacitance up to 304 Fg^{-1}.[9]

Graphene-conducting polymer hybrids also affect supercapacitor performance. Initially, the incorporation of GS can increase the electrical conductivity of polyaniline (PANI), and hence can raise power density. The GS can act as a conducting network in the nanocomposite system, which enhances the electrochemical redox reaction of PANI. Furthermore, both the charge storage processes comprising of electric double-layer charging/discharging of GS and pseudocapacitive redox reactions of PANI take place in the nanocomposite system that results in good functioning of supercapacitor. The PANI/graphene composite papers have been fabricated by anodic electropolymerization of 0.05 M aniline monomer on GS. This graphene-based nanocomposite exhibits stable electrochemical capacitance of 233 Fg^{-1}, which is represented in Figure 2.10. The electrochemical properties of graphene paper and PANI/graphene composite papers (Fig. 2.10A), cyclic voltammetry (CV) recorded from 2–20 mVs^{-1} in 1 M H_2SO_4 (Fig. 2.10B), Nyquist plots of graphene paper and PANI/graphene composite papers 60s/300s/900s (Fig. 2.10C) and cyclic performance measure at 50 mVs^{-1} (Fig. 2.10D) are represented in Figure 2.10.[7,30]

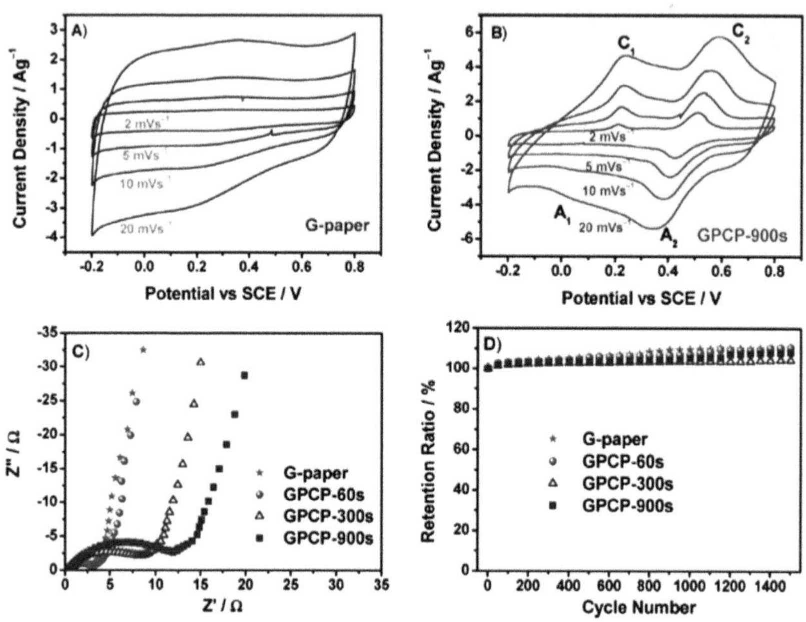

FIGURE 2.10 Electrochemical properties of PANI/graphene composite papers.[30]

Source: Reprinted with permission from Ref. [30]. © 2009 American Chemical Society.

In an alternative method to fabricate the composite electrode, GO is taken as nanofiller and PANI is in situ polymerized. This method results a large specific capacitance of 531 Fg^{1} at a weight ratio 100:1 (aniline/GO) whereas individual PANI shows 216 Fg^{1}. On the other hand, a nanocomposite prepared by uniformly coating GS with PANI matrix exhibits together with superior conductivity, yields large specific capacitance of 480 Fg^{-1} at a weight ratio 20:80 (aniline/GO). Additionally, the combination of 1D and 2D carbon nanomaterials can be used to modify the porous structures of the nanocomposite films and improve structural stability. Graphene/CNT nanocomposite sheets have been prepared by using layer-by-layer (LBL) assembly techniques, which possess consistent interconnected carbon nanostructures. The supercapacitor fabricated from this film displays a specific capacitance of 120 Fg^{-1}.[7]

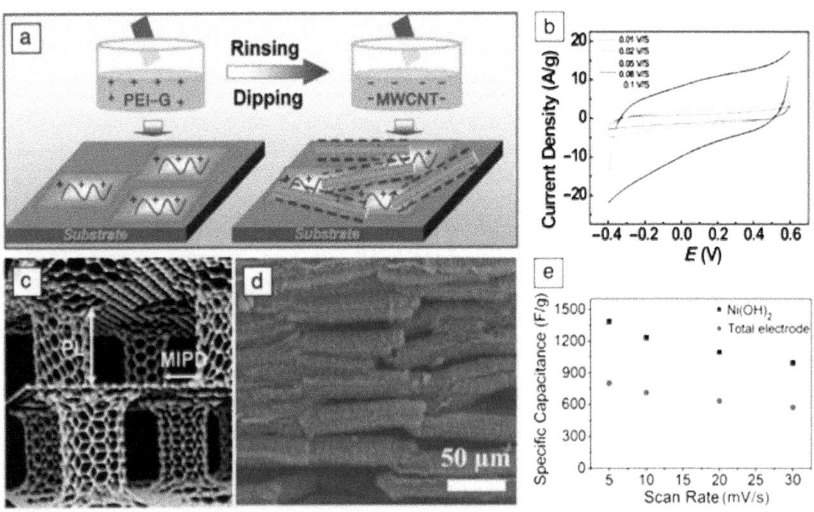

FIGURE 2.11 Schematic representation of supercapacitor with MWCNT and VA-CNT hybridized with graphene materials.[6]

Source: Reprinted with permission from Ref. [6]. © 2012 Springer Nature.

For example, a solution layer-by layer self-assembly method (shown in Fig. 2.11a and 2.11b) was applied to make multilayered composite films of polyethylene imine-(PEI) modified GS and acid oxidized multiwalled carbon nanotubes (MWCNTs) to construct supercapacitor having average specific capacitance 120,000 F/kg that is 120 Fg^{-1}. In the same way, the inherently nanoporous 3D pillared vertically aligned CNT (VA-CNT)/graphene composite structure with high surface area (Fig. 2.11c) can be beneficial for supercapacitance.[6] Therefore, Du et al. fabricated a VA-CNT/graphene composite by the intercalation synthesis of VA-CNTs into the thermally lengthened highly ordered pyrolytic graphite as shown in Figure 2.11d.[31] The resulting VA-CNT/graphene embedded with $Ni(OH)_2$ exhibits large specific capacitance with good charging rate efficiency as shown in Figure 2.11e.[6]

To avoid the aggregation of graphene, Wang et al. attained a large range of capacitance nearly 205 Fg^{-1} at 1.0 V in electrolytic solution with aqueous phase, from the process of gas–solid reduction.[26] To stop the rearrangements of sheets "Liu" arranged the supercapacitors of graphene in a curvy manner that obtained a noticeable large density of energy i.e. 85.6 Wkg^{-1} at 30°C and 136 Wkg^{-1} at 80°C via ionic liquid electrolytes.[32] Further studies suggested the use of CNT as a spacer between the GS stop agglomeration and thereby increasing the specific capacitance up to 326 Fg^{-1} with a proportion of 1:9 for the weight of CNT to graphene sheet, respectively. Another work suggests that though Iron (II/III) oxide/ Fe_3O_4 has low electrical conductivity, but the sheets (reduced form of GO) with composites of Iron (II/III) oxide nanoparticles (NPs) exhibit a capacitance nearly about 480 Fg^{-1}.[22]

Graphene component can offer pliable electrode materials having extremely conductive platform for the transfer of charge and also have considerable surface areas for the contact of electrolyte, otherwise, it deactivates rest of the functional material. Those tuff mechanical properties of graphene-based molecules also help in the stability process and flexibility process of supercapacitors. Those Graphene electrodes might be fabricated with deposition of graphene molecules on pliable substrates, such as textiles, sponges, papers. Graphene electrodes were prepared by simple coating of GO inks on the surface of cotton cloth and then go through annealing treatment. Those flexible electrodes exhibit a specific capacitance at about 81.7 Fg^{-1}. In this manner, the composites of cellulose paper or graphene have been fabricated. The fabrication process involves filtering, then corresponding interruption of RGO sheets into the holes of filter papers. For which, the flexible supercapacitor based on this electrode shows a specific capacitance of about 120 Fg^{-1} determined by CV curve at 1 mVs^{-1} in 1 M H_2SO_4 aqueous electrolyte.[33]

Graphene aerogel containing high porous volume increases the connection area between the GS and electrolytes.[34] Zhang and Shi prepared graphene hydrogels using hydrothermal process. The capacitance value determined is 160 Fg^{-1}, which is credited to the porous 3D network structures of graphene. Then, the hydrogel is reduced by N_2H_4 that leads to increase in capacitance of 222 Fg^{-1}.[34,35] Luan et al. synthesized graphene aerogels that cross-linked with ethylene diamine, which are reduced to produce conductivity of 13.51 Scm^{-1} and specific capacitance of 231 Fg^{-1}.[34,36]

Choi et al. fabricated graphene/PANI multilayered nanostructure (GPMN) as active electrode material using simple sonication method. The GPMN sheet on a filter membrane shows outstanding flexibility. The material exhibits a specific capacitance of 390 Fg^{-1} upon dispersing GPMN in water (GPMN-W) with mass ratio of GPMN-W 1:60 in comparison with GPMN dissolved in N-methyl-2-pyrrolidone (GPMN-N).[37,38] Similarly, Xiao et al. also prepared a sandwiched structure of graphene/PANI/graphene nanocomposite film via a scalable and modular process. Incorporation of PANI nanofibers into the nanocomposite improves energy storage capability with electrical conductivity of 340 Scm^{-1} and specific capacitance of 581 Fg^{-1}.[37,39]

For the construction of composites of metal oxide or graphene (Being used for making supercapacitors) mainly three-dimensional networks of graphene are being used. Due to long-range specific capacitance, adequate reducing property and affordable price metal oxide like oxide of nickel (NiO) is fluently used in supercapacitors. Both the electrical and chemical activities of electrode made up of NiO/graphene composite were examined by CV curve and charge–discharge method using 3 $molL^{-1}$ potassium hydroxide aqueous solution in the voltage from 0 V to 0.5 V. The cyclic voltammetry curve of NiO/graphene at several scan rates, that is, from 5–40 mVs^{-1} is shown in Figure 2.12a. The current response is improved with the increase of scan rate, but such noticeable change in curve was not observed which further indicates that the electrode of NiO/graphene composite has admirable rate and property. At a scan rate of 5 and 40 mVs^{-1}, NiO/graphene nanocomposite illustrates specific capacitance of 816 Fg^{-1} and 573 Fg^{-1}, respectively (calculated from cyclic voltammeter curve). The Galvano static discharge plots of NiO/graphene composite at current density range of 1.4–20 Ag^{-1} is shown in Figure 2.12c. Cyclic activity of NiO/graphene composite was observed at a scan rate of 80 mVs^{-1} as displayed in Figure 2.12d. Specific capacitance improved by 15% in the course of 200 cycles.[40]

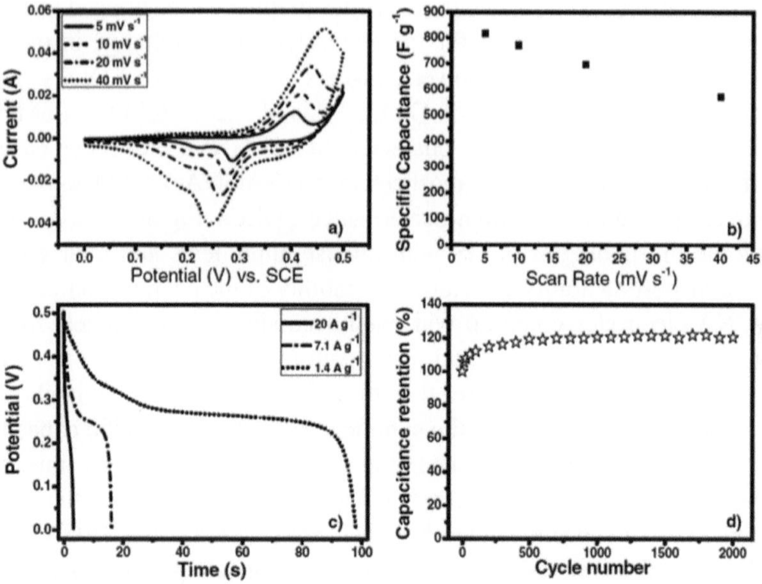

FIGURE 2.12 Electrochemical performance of NiO/graphene nanocomposite.[40]
Source: Reprinted with permission from Ref. [40]. © 2011 John Wiley.

The aminated graphene honeycomb (AGH) structure is synthesized via thermal expansion of graphitic oxide and then amination reaction. At low-temperature amination reaction (for example 200°C), NH_3 condensed with –COOH groups to generate generally $–CONH_2$ or $–NH_2$ groups (chemical N) by means of nucleophilic substitution reaction. Alternatively, at high-temperature amination reaction, the intramolecular dehydration or decarbonylation occurs to produce thermally stable N-based heterocyclic aromatic species likely C_5H_5N, C_4H_5N, and quaternary type N molecules (lattice N). This functional 3D assembly of graphene exhibits excellent energy storage capability with high capacitance (e.g., specific capacitance 0.84 Fm^{-2} and gravimetric capacitance 207 Fg^{-1}).[41]

Due to the presence of more Vander wall forces of attraction in-between the GO nanosheets (packed horizontally within a substrate), those sheets are continuously aggregating, for which it hinders the maximal uses of the surfaces. Therefore, to overcome this problem, it is necessary to align GS in an ordered manner, which can show effective utilization of the surface. For instance, an electrode comprising of vertically aligned GO sheets was fabricated by coating a GO layer into a bar and slicing along its longitudinal direction. It results in an increase in the graphene's packing density and availability of shortest path for the electrolytic ion exchange in electrode and thereby increasing the power densities and energy of EDLCs. This is represented in Figure 2.13(a–c). On the basis of successful fusion of unique structural design and intrinsic characteristics of GNSs, electrodes based on vertically aligned graphene nanosheet arrays (VAGNAs) could convenience the overall conditions for high-performance EDLCs. Figure 2.13(d–i) represents the morphologies of GS with diverse alignments and their effect on ion's transporting ability and storage capability of the charge. Comparatively VAGNA electrodes come up with well accessible open channels for electrolyte and provide simplest and linear path for exchanging of ions and also increases the use of surface area than that of graphene layers having horizontally packed. This concludes the significant increase in specific capacitance of 38% more than that of the lateral graphene electrode and six times more capacitance than that of the graphite sheet (Fig. 2.13g). From the Nyquist plot and rate of capacity data, it has been confirmed that the VAGNAs electrode showed lower resistance and the finest capacitance retention of 82% in comparison to other two electrodes, which is presented in Figure 2.13h and 2.13i. For pseudocapacitors, the function of VAGNA electrodes can be additionally enhanced by involving transition metal oxides that stock charge via rapid and extremely reversible Faradaic redox reactions on the surface.[42]

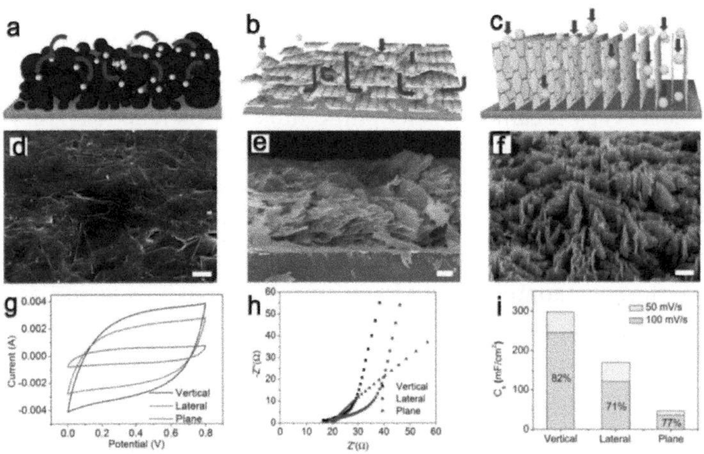

FIGURE 2.13 Schematic representation, SEM images, and electrochemical performance of graphene powder, vertically aligned graphene, and horizontally aligned graphene.[42]

Source: Reprinted with permission from Ref. [42]. © 2017 John Wiley.

Polymer/graphene nanocomposites are useful in supercapacitor application. Polyaniline/graphene nanocomposites are easy to fabricate electrochemically or chemically and they can show capacitances ranging from 233–1046 Fg^{-1} depending upon the nanostructures of composites. Supercapacitor comprises of polysodium 4-styrensulfonate (PSS)-graphene nanocomposite showed very high cycle stability. The specific capacitance value 190 Fg^{-1}, reduced by 12% after 14860 cycles. Graphene or composites of polymer are preferably appropriate for transportable and wearable electronics as they are pliable and their characteristics remain unchanged though they are under mechanical pressure. For instance, Wu et al. prepared pliable composite of polyaniline/graphene with capacitance value of 210 Fg^{-1} and cyclic stability of 94%.[1,43]

Choi et al. prepared a solid-state supportive pliable supercapacitor by arranging a sheet electrode of Nafion functionalized RGO (f-RGO) and solvent cast Nafion electrolyte membranes as electrolyte and separator. This f-RGO film showed high flexibility and the as prepared supercapacitor showed good capacitance of 118.5 Fg^{-1} at 1 Ag^{-1}.[44,45]

Liu et al. organized a "dipping and drying" scheme to overlay the cellulose fibers with a fine and consistence GO layer and then accumulated GO films into RGO networks to fabricate RGO/cellulose fibers nanocomposite sheet, which contains plentiful microsized holes.[46] Those structure-oriented composite concurrently alleviates the accumulation of RGO in holes of paper and encouraging the diffusion process of ions in electrolyte by completely employing the cellulose fibers network and the porosity of the paper. Further, there was deposition of PANI on the surface of RGO to attain a long-range capacitance of 464 Fg^{-1}. The PANI-RGO/CF composite papers possess outstanding flexibility and conductivity that allow the direct utilization as electrodes for flexible supercapacitors with thin Pt wire as the current collector.[44,46]

Cong et al. introduced a suitable single-step process to synthesize freestanding, graphene paper having lower density with a decent electrical conductivity, where PANI nanorod arrays may be electrochemically grown to develop pliable paper of graphene–PANI composite. As prepared composite electrode shows high flexibility and exhibits an increase in the activity of capacitor, that is, the capacitance value increases up to 763 Fg^{-1} at a discharge current of 1 Ag^{-1} and also shows considerable cycling stability with efficiency of 82% capacity retention after 1000 cycles.[47] In addition to polyaniline (PANI), polypyrrole (PPy) also used for the improvement of capacitance of pliable graphene electrodes.[44,47] Lu et al. suggested a flexible graphene-PPy/CNT ternary nanocomposite via a process of flow-assembly, where there was consistently sandwiching of coaxial PPy/CNT nanocables within GS, for which the mechanical stability of PPy chains was well enhanced by pliable GS and the hard CNT core. The specific capacitance of the G-PPy/CNT nanocomposite was recorded as 211 Fg^{-1} at a discharge current of 0.2 Ag^{-1} and good cycling stability of 5% capacity loss after 5000 cycles was observed.[44,48] He et al. proposed a kind of MnO_2-coated freestanding, pliable, thin 3D graphene network having high conductivity, on which an expected large weight content of 92.9% MnO_2 was observed. This pliable electrode shows capacitance of 1.42 Fcm^{-2} and large specific capacitance of 130 Fg^{-1}, and it is interesting to know that those electrodes could be twisted largely without any performance loss.[44,49]

2.4.2 LITHIUM-ION BATTERY (LIB)

This type of battery comprises of mainly three key constituents, such as cathode, anode, and Li-ion conducting electrolyte. LIB is a secondary or rechargeable battery, works by transforming a chemical energy into electrical energy through Faradic reactions, that is, it involves the process of heterogeneous charge transfer, taking place at the electrodic surface. The charging process of LIB includes a reverse of energy from electrical energy to chemical potential. Faradic reactions in LIBs go with weight as well as charge transfer through electrodes and dimensional variation, thus the distance of migration and area under surface are the crucial factors which predict the activities of batteries. The configuration, structure of crystal, and morphology can reveal the rate of reaction, process for transferring of electrode material and also can be deployed to change completely the electrochemical property. Graphene shows high conductivity, high mechanical strength and large surface area. Therefore, many 1D nanostructured materials, such as SnO_2, MnO_2, TiO_2, Co_3O_4, Cu_2O, CNTs, carbon fibers have been manufactured with graphene for high-performance LIBs.[22,50]

It is observed that graphitic carbon can form LiC_6 structures. Graphitic carbon is extensively utilized as an anodic material in LIBs, whereas somewhat the lowering in density of lithium in graphite causes the corresponding decrease of specific capacity of graphite that is 372 $mAhg^{-1}$. When lithium is being stored on one and the other sides of graphene sheet, it forms structures of LiC_3, and then by engaging individual GS, the storage capacity limit becomes 744 $mAhg^{-1}$. Stacked platelet graphene nanofibers show excellent electrochemical properties as their surfaces show virtually edge-like sites and interlayer spacing of the graphite, whereas basal planes are positioned at the termination side of nanofibers. Such materials show increased energy storage capacity of 461 $mAhg^{-1}$. Therefore, it is clear that lesser the stacked graphene platelet nanofiber, greater will be the energy-reserving capacity.[1] Figure 2.14 represents the working principle of Li-ion battery.

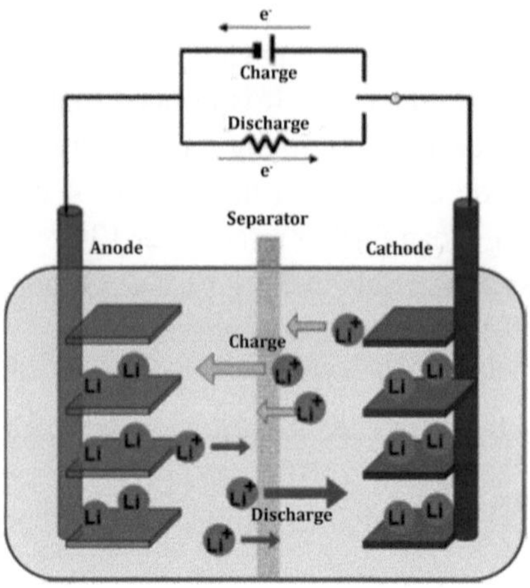

FIGURE 2.14 Schematic representation of working principle of li-ion battery.[51]
Source: Reprinted with permission from Ref. [51]. © 2012 John Wiley.

Yoo et al. worked on various possibilities to increase the lithium storage capacity by GNS material. They managed the layered form of GNSs through the process of reassemble and exfoliation. Then the specific capacity observed was 540 $mAhg^{-1}$ and which was promoted to 730 $mAhg^{1}$ and 784 $mAhg^{-1}$, respectively by employing CNT and C_{60} to GNS.[52] Relationship within the interlayer spacing and the charge capacity of GNS and graphite is shown in Figure 2.15A while cross-sectional TEM microphotographs of GNS with almost same quantities (5–6) of graphene stacking layers for (a) GNS, (b) GNS+CNT, (c) GNS+C_{60} are presented in Figure 2.15B.[52]

Wu et al. reported an electrode that was built by heteroatom (N, B)-doped graphene (derived chemically) that exhibited extremely high charge–discharge rate and high capacity. The doped electrodes exhibited the capacity of 1043 $mAhg^{-1}$ and 1540 $mAhg^{-1}$ for N-doped and B-doped graphene, respectively at a small charge/discharge rate of 50 mAg^{-1}.[53] It was observed that the rGO/Fe_3O_4 nanocomposite exhibited excellent capability and stable cyclability compared with that of commercial Fe_3O_4 and Fe_2O_3 particles. Specifically, the specific capacity of the rGO/Fe_3O_4 nanocomposite is raised to 520 $mAhg^{-1}$ when the current density reached 1750 mAg^{-1}. Similarly, in LIB, the Mn_3O_4/rGO composite, as anode exhibited higher rate capability as compared with bare Mn_3O_4 NPs. The specific capacity of the Mn_3O_4/rGO was less than 390 $mAhg^{-1}$ at the current density of 1600 mAg^{-1} that is also relatively larger in comparison with the specific capacity of graphite. Again, rGO-encapsulated Co_3O_4, denoted as rGO@Co_3O_4, demonstrated higher capacity and excellent cycling stability to that of assorted Co_3O_4/rGO composite or pure form of Co_3O_4. It is noticed that an extremely reversible specific capacity of nearly 1100 $mAhg^{-1}$ transferred initial 10 cycles of the rGO@Co_3O_4 electrode at a current density of 74 mAg^{-1}, and a discharge capacity of 1000 mhg^{-1} can further retain after 130 cycles. In addition to this, some graphene-based materials like rGO/$LiFePO_4$ composite can be used as cathode in LIBs and can enhance the capacity.[2]

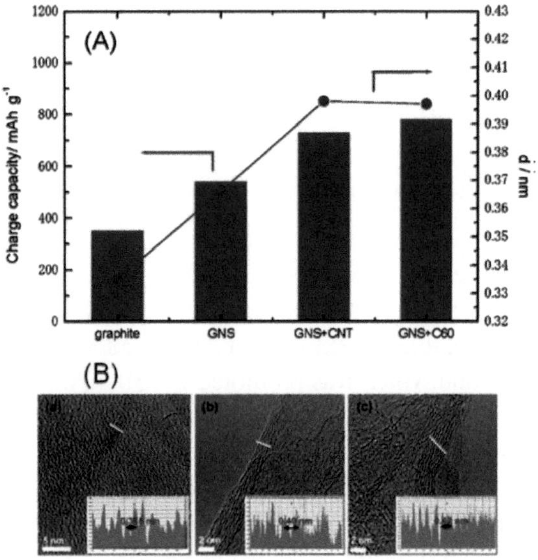

FIGURE 2.15 (A) Charge storage capacity and (B) TEM microphotographs of graphene-based materials.[52]

Source: Reprinted with permission from Ref. [52]. © 2008 American Chemical Society.

It was observed that there was existence of higher reversible capacity to that of bare graphite electrode, bare SnO_2 and graphene, for graphene/SnO_2-based nanoporous electrode. The graphene/SnO_2-based nanoporous electrode showed a reversible capacity of 810 $mAhg^{-1}$. Further, the cyclic performance was extremely increased as compared with bare SnO_2 NPs where after 30 cycles, the charge capacity of graphene/SnO_2-based electrode remained at 570 $mAhg^{-1}$ while the first charge capacity of SnO_2 NPs was 550 $mAhg^{1}$ that has fallen quickly to 60 $mAhg^{-1}$ after only 15 cycles at 50 mAg^{-1}. The structure and synthesis of graphene-SnO_2 electrode is presented in Figure 2.16A. Figure 2.16B demonstrates the capacities and cyclic abilities of (a) SnO_2 NP, (b) graphite, (c) GNS, (d) SnO_2/GNS, and the improved lithium storage-release cyclability of graphene–SnO_2 electrode is shown in Figure 2.16C.[6,54,55] To assimilate the complete use of both graphene and metal oxide, there was development of various nanocomposites for anodic materials. Although oxide of metal can possess larger lithium-ion storage capacity, this anode has a major disadvantage that a large change of volume at the time of charge/discharge method ensuring in cracking of electrode and low reversibility. In nanocomposites, graphene can perform as both mechanical support layer to alleviate the cracking and strong conducting layer for charge carriers.[51]

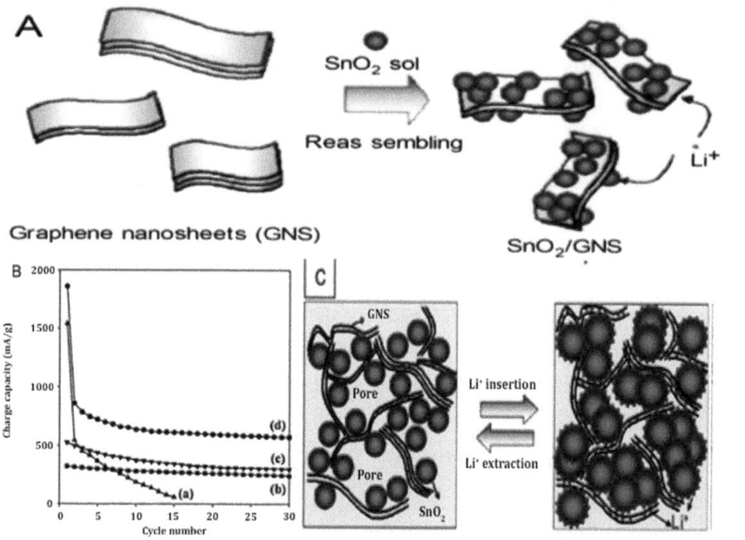

FIGURE 2.16 Schematic representation of graphene/SnO_2 hybridized electrode.[6,54]

Source: Reprinted with permission from (a,c) Ref. [6]. © 2012 Springer Nature; (b) Ref. [54]. © 2011 Elsevier.

Yang et al. proposed a chemical method to produce 2D sandwiched type graphene-based mesoporous silica (GM-silica) sheets, that is, there was complete separation of graphene sheets by a shell of mesoporous silica.[56] The sandwich as sheets of GM-silica were utilized as an arrangement to prepare graphene-based mesoporous carbon (GM-C) and Co_3O_4 (GM-Co_3O_4) films. Storing capacity of Li ion within GM-C sheets was calculated and an extreme reversible capacity of 915 mAhg^{-1} was obtained at a rate of C/5. After 30 cycles, both discharge/charge capacities of GM-C films become constant at around 770 mAhg^{-1} and showing 84% capacity retention. This supports the fact that graphene-based nanosheets can increase the electrochemical performance.[9]

The rGO films fabricated by thermal pyrolysis at 300°C displayed high reversible capacity of 794–1054 mAhg^{1} and thereby conclude that the incorporation of large quantity of crystal defects in graphene was advantageous to increase the lithium-ion storage capacity.[11] Fang et al. fabricated mesoporous GNSs by controllable low concentration monomicelle close-packing assembly method. LIBs anodes prepared with these mesoporous GNSs displayed a maximum reversible capacity of 1040 mAhg^{-1} at 100 mAg^{-1}, and even after dozens of cycles with different current densities, they could retain 833 mAhg^{-1}.[11,57] In another approach, Yin et al. synthesized GNSs into a 3D honeycomb structure, which showed high conductivity. This anode showed a high reversible lithium ions storage capacity of about 1600 mAhg^{-1}, and after 50 cycles, showed stable cycling performance of 1150 mAhg^{-1}.[11,58]

Graphene aerogels are also used as cathodes, anodes in lithium-ion batteries. $LiFePO_4$ (LFP) has gained attraction as cathode materials owing to its various interesting characteristics, such as adequate stability, strong capacity, cheap and environment-friendly properties.[34] Wang et al. mixed N-doped graphene aerogels with LFP, which minimize the diffusion length of lithium ion in LFP crystals, and ease availability of fast route for transport of lithium ion. A cathode made up of LFP improved by 3D graphene showed outstanding stability and high-rate capability of 78 mAhg^{-1} at 100°C.[59] In another work, Wang et al. synthesized porous aerogel/O_2 cathodes for Li-O_2 batteries. The batteries acquired the capacities of 11060 mAhg^{-1} at 0.2 mAcm^{-2} (280 mAg^{-1}) and 2020 mAhg^{-1} at 2 mAcm^{-2}.[60] Ren and his co-authors prepared Li–S batteries with maximal sulfur loading on graphene aerogel (GA) which had a capacitance value of 1000 mAhg^{-1}.[61] Feng also prepared an interlinked microporous 3D

Fe_3O_2/graphene aerogel, which exhibited a maximum reversible capacity of 372 $mAhg^{-1}$ at high rate of 5000 mAg^{-1} and 995 $mAhg^{-1}$ after 50 cycles at 100 mAg^{-1}.[62] In this manner, a TiO_2/graphene aerogel demonstrated a capacity of 200 $mAhg^{-1}$ at 0.59 C after 50 cycles and a capacity of 99 $mAhg^{-1}$ at 5000 mAg^{-1}. Again, wrapped Bi_2O_3/graphene aerogel showed a high capacitance of 417 $mAhg^{-1}$ over 100 cycles at 200 mAg^{-1} and 273 $mAhg^{-1}$ at 10000 mAg^{-1}.[34] Wei et al. prepared Fe_3O_4/GA that showed a capacity of 63 $mAhg^{-1}$ at 4800 mAg^{-1} and cycling performance of 1059 $mAhg^{-1}$ over 150 cycles at 93 mAg^{-1}.[63] Garakani et al. gave an idea of improved charge carrier for enhanced cycle stability in a Co_3O_4/GA anode which exhibited following specific capacities of 1001, 900, and 798 $mAhg^{-1}$ at 0.2, 0.5, and 1.0 Ag^{-1}.[64] Meng et al. produced amorphous SiO_2/GA anodes that showed stable cyclic performance and reversible capacities of 300 $mAhg^{-1}$ at 500 mAg^{-1}.[65]

Many polymers, such as sodium carboxy methyl cellulose (SCMC), polyimide, phenol formaldehyde resin (PF), polyaryl triazine were employed into graphene to prepare the anode electrodes of LIBs.[37] Jeong et al. fabricate a monolithic graphene as LIB anodic material via freeze-drying of a mixture of rGO and SCMC. The monolithic graphene has great surface area, high electrical conductivities and short diffusion lengths that bring about superfast charge/discharge rates of 718 $mAhg^{-1}$ at 5 C (12 min.) and 550 $mAhg^{-1}$ at 100 C (36 s).[66] Li et al. proposed PF-grafted rGO (rGO-g-PF) as anodic material for LIBs. After coating of PF and a carbonization process, rGO-g-PF shows a porous structural system that can successfully detain co-interaction of solvated lithium ions and exfoliation of graphene layers that further leads to huge lithium-ion storage capacity. rGO-g-PF as anodic material demonstrates remarkably good rate capability (301.9, 337.9, 362.8, and 400.8 $mAhg^{-1}$ at 0.5, 0.2, 0.1, and 50 mAg^{-1}, respectively) and high cyclic stability (376.5 $mAhg^{-1}$ at 50 mAg^{-1} for 250 cycles, 337.8 $mAhg^{-1}$ at 200 mAg^{-1} and 267.8 $mAhg^{-1}$ at 1 Ag^{-1} for 200 cycles) which convert it into a good anode electrode material for LIBs.[67] Su et al. have fabricated 2D-coupled graphene/porous polyaryl triazine-based frameworks as a LIB cathode and as prepared LIB system exhibits an outstanding rate capability of 135 $mAhg^{-1}$ at a high current density of 15 Ag^{-1} and outstanding cycle stability of 395 $mAhg^{-1}$ at 5 Ag^{-1} for 5100 cycles.[68]

The minimum theoretical specific capacity of the graphene restricts the direct application of VAGNA electrodes in LIBs. However, the binder-free

VAGNA electrodes displayed reduced electrical resistance for transport of charge and thus increased in kinetics. The VAGNAs grown on Cu surface act as an anode for LIBs with specific capacities of 500 $mAhg^{-1}$ and 297 $mAhg^{-1}$ at a current density of C/5 and 4C, respectively, with 10% reversible capacity loss after 100 cycles. By comparing the binder-free horizontally packed graphene nanoflakes synthesized by the process of exfoliation on Cu surface, the VAGNA demonstrated lower capacity at lower current density (500 $mAhg^{-1}$ vs 610 $mAhg^{-1}$ at ~C/5) but showed comparatively a better capacitance of (297 $mAhg^{-1}$ vs 150 $mAhg^{-1}$ at ~4C). In addition to this, the Sn/VAGNA anodes exhibited a reversible capacity of 466 $mAhg^{-1}$ at a current density of 879 mAg^{-1} after 4000 cycles, and 794 $mAhg^{-1}$ at 293 mAg^{-1} after 400 cycles. When VAGNAs with hydrothermally coated ZnO NPs were utilized as electrodes, they delivered a specific capacity of 810 $mAhg^{-1}$ at a current density of 80 mAg^{-1}. Moreover, when amorphous GeO_x layer was set down on the surface of VAGNAs, it exhibited a firm capacity higher than 1000 $mAhg^{-1}$ within 100 cycles at C/3 and a long-range capacity of 545 $mAhg^{-1}$ at an improved rate of 15 C. Additionally, when Si NPs popped on surface of VAGNAs substrate were used, it provides a long-range capacity of ~2000 $mAhg^{-1}$ at 1.5 C rate after 40 cycles. When Si thin films were deposited on VAGNAs, particularly, Si/3D-graphene composite used, it displayed a gravimetric capacity of 1314 $mAhg^{-1}$ at C/5 after 500 cycles with much better capacity retention than Si NPs on horizontal graphene. When the MoS_2/VAGNA/Ni-foam anode used, it provides a high specific capacity of 1277 $mAhg^{-1}$ at current density of 100 mAg^{-1} and high capacity of 818 $mAhg^{-1}$ at a raised current density of 2000 mAg^{-1}.[42]

A novel tactic was proposed to design and synthesize a ternary rGO/Fe_2O_3/SnO_2 graphene nanocomposite which consists rGO with integration of highly Li active Fe_2O_3 and SnO_2 particles. The synthesis process of rGO/Fe_2O_3/SnO_2 ternary composite via in situ precipitation of Fe_2O_3 NPs onto GO surface, which is subsequently reduced with $SnCl_2$ to get conducting rGO is represented in Figure 2.17a. Assessment of discharge/charge capacity calculated with rGO, rGO/SnO_2 and rGO/Fe_2O_3/SnO_2 anodes up to 100 cycles is presented in Figure 2.17b.[69] As represented in Figure 2.17, the rGO/Fe_2O_3/SnO_2 electrode is better than the rGO and rGO/SnO_2 anodes, resulting in enhanced cycle stability and specific capacity. After 100 cycles, the charge capacities of rGO/Fe_2O_3/SnO_2 anode are the highest as compared with rGO and rGO/SnO_2, and maintaining a specific capacity over 700 mAhg-1 that can be maintained during subsequent 100 cycles. In this rGO/Fe_2O_3/SnO_2 ternary anodic material, rGO, Fe_2O_3, and SnO_2 particles act differently but show complementary functions. Therefore, rGO behaves as a matrix permitting both lithium ions and electrons to transfer to active sites, thus completely increasing the energy density tendered by both graphene and Fe_2O_3. The rGO also is an efficient elastic buffer to release strain or else collected in the agglomerated Fe_2O_3 particles during lithium-ion uptake/release. Again, the dispersion of Fe_2O_3 and SnO_2 NPs on graphene stops the restacking of GS, retaining a high storage capacity of lithium ions during cycling.[69]

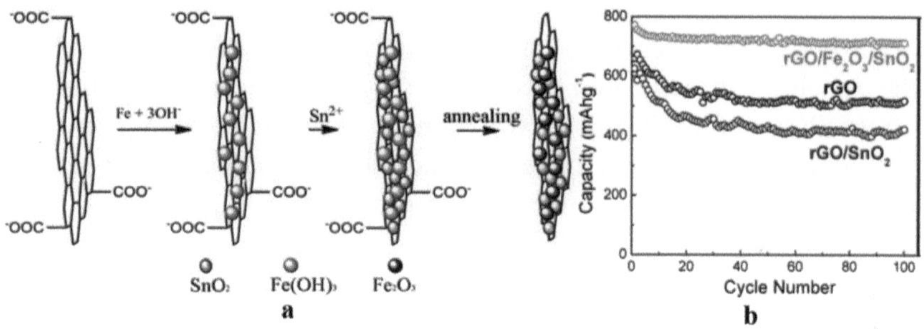

FIGURE 2.17 Schematic representation of (a) synthesis of rGO/Fe_2O_3/SnO_2 ternary nanocomposite and (b) charge storage capacity of rGO and rGO/SnO_2, and rGO/Fe_2O_3/SnO_2 ternary nanocomposite.[70]

Source: Reprinted with permission from a) Ref. [70]. © 2014 American Chemical Society.

The edge selectively hydrogenated and halogenated GNSs fabricated by ball milling method are potential materials as anode for high-performance LIBs. For instance, edge iodinated graphene-based LIBs were observed to produce an initial charge capacity of 562.8 mAhg-1 at 0.5 C in a potential range of 0.02–3.0 V. Further, these LIBs showed a useful cycling stability (charge capacity retention of 81.4% after 500 cycles) as well as a decent long-term life (high reversible capacity of 461.1 mAhg1 after 30 days of storage). On the other hand, edge-fluorinated graphene nanoplatelets showed a charge capacity of 650.3 mAhg-1 at 0.5 C with charge retention of 76.6% after 500 cycles.[71]

The few-layer MoS_2/GNS and SnS_2 nanocomposites were prepared by L-cysteine supported hydrothermal synthesis. These two nanocomposites produced a reversible specific capacity of around 900–1200 $mAhg^{-1}$ with great cycle stability and improved rate capability. Exfoliated MoS_2/GNS/PEO nanocomposite delivered high capacity of 1080 $mAhg^{-1}$ and improved rate property. A 3D porous MoS_2/GNS structural design fabricated by hydrothermal process and followed by freeze-drying displayed good specific capacity value of 1200 $mAhg^{-1}$ with increased rate capability and long cycle-life of 3000 cycles.[72]

2.5 GRAPHENE-BASED MATERIALS FOR ENERGY CONVERSION

2.5.1 SOLAR CELL

Solar cells also referred as photovoltaic cells are devices that change the sun light directly into electrical energy by means of photovoltaic effect. Polymer solar cells (PSCs) consist of an active layer consists of both donor and acceptor materials enclosed among cathode and anode with a condition that one of the electrodes must be transparent to accept sunlight. The layer-by-layer stacked graphene grown on Cu substrate and doped with acid to give carriers was demonstrated to show a sheet resistance of 80 Ω/sq and a transmittance of 90% at 550 nm. PSCs containing this layered graphene electrode and MoO_3 as a hole extraction layer exhibited a PCE of 2.5%. GO derivatives were developed to use as hole extraction and electron extraction layers in PSCs. GO was used as hole extraction layer to achieve a power conversion efficiency (PCE) in comparison with the poly(3,4-ethylenedioxythiophene):poly (styrene sulfonate) (PEDOT:PSS) layer. Alternatively, cesium-doped GO (GO-Cs) which prepared using charge neutralization of the peripheral –COOH groups of GO with Cs_2CO_3, work as significant electron extraction layer. PSCs containing GO and GO-Cs as hole extraction and electron extraction layer, respectively, displayed a PCE of 3.67% that is higher as compared with advanced device with hole extraction and electron extraction layer, that is, 3.15%.

Yu et al. prepared chemically grafted –CH_2–OH terminated regioregular poly(3-hexylthiophene) (P3HT) onto –COOH groups of GO by means of esterification reaction to yield P3HT-grafted GS (G-P3HT).[73] A bilayer photovoltaic device prepared from solution cast C_{60}-G:P3HT displayed PCE value of 0.61%, which was twice of its C_{60}:P3HT counterpart. The device structure of PSC in normal configuration is shown in Figure 2.18a, while the device structure of normal (left) and inverted (right) PSCs with GO as hole extraction and GO-Cs as electron extraction layer is presented in Figure 2.18b. In addition, the C_{60} grafted graphene (C_{60}-G) and its PSC device functioning is presented in Figure 2.18c and 18d, respectively.[6]

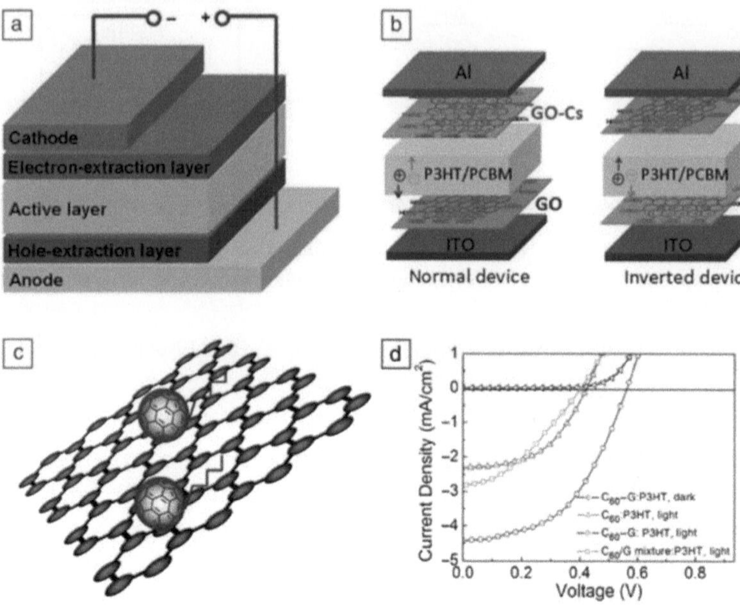

FIGURE 2.18 Schematic representation of device structure of PSC containing graphene-based materials and performance studies.[6]

Source: Reprinted with permission from Ref. [6]. © 2012 Springer Nature.

The structure and synthesis pathway to GO-Cs are shown in Figure 2.19a. Similarly, Figure 2.19b demonstrates the synthetic method for chemical grafting of $–CH_2–OH$ terminated P3HT onto the graphene layers that includes $SOCl_2$ treatment of GO in step 1 and esterification reaction between acyl chloride functionalized GO and Me–OH-terminated P3HT in step 2. The energy level diagrams of normal device (left) and inverted device (right) with GO as the hole extraction layer and GO-Cs as the electron extraction layer are shown in Figure 2.19c and J–V curves of normal and inverted device are represented in Figure 2.19d and 19e, respectively.[51,19]

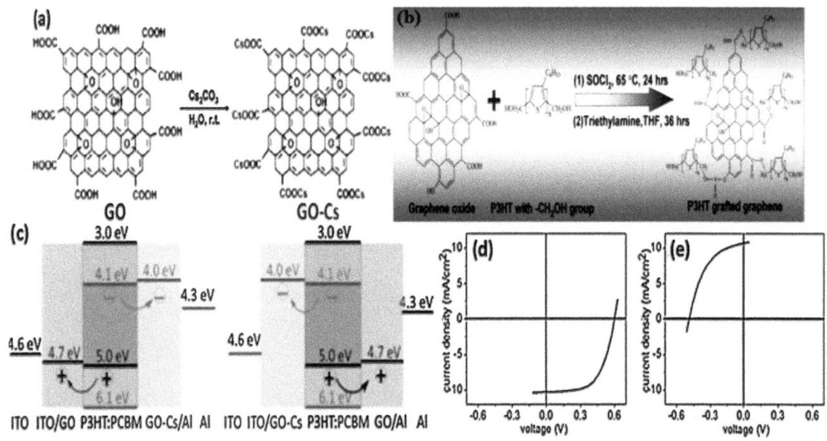

FIGURE 2.19 Schematic representation of synthesis of (a) GO-Cs[19] and (b) P3HT-grafted graphene[51] and (c, d, and e) energy level diagram with J–V curve of normal and inverted device.[19]

Source: Reprinted with permission from (a) Ref. [19]. © 2016 Elsevier; (b) Ref. [51]. © 2012 John Wiley.

Various solar cells based on graphene material were proposed, among them the graphene acts as another vital portion of the cell. Photovoltaic devices have been constructed by using CVD graphene as transparent conductive electrodes. The prepared device having a structure of graphene/polyethylene dioxythiophene doped with polystyrene sulfonic acid (PEDOT:PSS)/poly(3-hexylthiophene):phenyl-C_{61}-butyric acid methyl ester (P3HT:PCBM)/LiF/Al exhibited an enhanced PCE of 1.71% when the hydrophilicity of graphene sheets was improved by noncovalent moderation with pyrene butanoic acid succidymidyl ester. The graphene sheets prepared by CVD were also used in solar cells with a structure of CVD graphene/PEDOT/copper phthalocyanine (CuPc)/fullerene (C_{60})/ bathocuproine (BCP)/Al and having a PCE of 1.18% recorded.[9]

Generally, PV devices based on p–n junctions considered as first-generation solar cells having efficiency about ~25%. To increase the efficiency, the development of second-generation PVs based on thin-film technology has been applied. The third-generation PVs depend on the exploitation of evolving organic PVs, dye-sensitized solar cells (DSSCs), quantum dot solar cells (QDSCs) that may be low-cost one, more versatile, and environment-friendly. Graphene-related materials (GRMs) are being developed as transparent conductors (TCs), photosensitizers, channels for charge transport, and catalysts. The performance/cost ratio can be improved by using GRMs for TCs to substitute indium tin oxide (ITO) and catalysts to replace platinum. For instance, the implementation of graphene as electrocatalysts for polypyridine complexes of Co(III/II) in DSSCs results in an efficiency value of 13%.[18] Typically, DSSC contains a transparent cathode (fluorine-doped tin oxide (FTO)), a highly porous semiconductor layer (TiO_2) with a soaked layer of dye (ruthenium polypyridine), an electrolyte solution containing redox pairs (iodide/triiodide) and a counter electrode (Pt). DSSCs are not effective in comparison with silicon solar cells, but their cost-effectiveness and simple preparation method make them potential for low-density application like rooftop solar collections.[6]

A novel method, which is prepared from a simple reaction between Li_2O and CO, was employed to prepare 3D honeycomb-like structured GS. This DSSC with honeycomb-structured graphene counter electrode showed energy-conversion efficiency of 7.8% that is equivalent to DSSC with a Pt counter electrode.[74] The schematic representation of device structure of DSSC and the use of GRMs in it is presented in Figure 2.20a and 2.20b, respectively.

Xue et al. fabricated N-doped graphene aerogel as a metal-free counter electrode for DSSCs with PCE up to 7.07%[75] while Ahn et al. prepared B-doped graphene aerogel which showed PCE up to 8.46%.[76] Yuan et al. altered polyacrylic acid/cetyl trimethyl ammonium bromide (PAA-CTAB) gel electrolytes with graphene to form a quasi-solid state DSSC. This enhances the strength of solid electrolytes with quick movement of charge within graphene to result PCE of 7.06% as compared with 6.07% for pure PAA-CTAB.[77] Kim et al. fabricated a TiO_2-embedded graphene photoanode having 3D porous structure, which exhibited photon to electric transformation efficiency of 7.5% with 55% improvement without the graphene in the system.[78] Ergen et al. have prepared the double-perovskite layers (where layer 1: $CH_3NH_3SnI_3$ and layer 2: $CH_3NH_3PbI_{3-x}Br_x$) by integrating GaN, hexagonal boron nitride (h-BN) and graphene aerogel. The cells showed average PCE of 18.41%.[34,79]

Well-organized structures of graphene molecules have been extensively considered as transparent films for the conduction of charge and also as catalytic counter electrodes (CEs) like a photovoltaic device because of excessive transparency to light and also in-plane conductivity of graphene layers and the catalytic property of doped graphene sheets.[44] Additionally, Guo and his co-author proposed the electron-accepting behavior of graphene in solar cell made up of CdS quantum dot (QD).[81] Author Liu described the uses of graphene (functionalized by isocyanate ion in solution), that is, it behaves as an electron acceptor in the solar cells having bulk heterojunction (BHJ).[82] The larger periphery of GQDs can suitably provide ample interfaces for exciton dissociation, and top movement of electrons of GQDs is anticipated to enhance the conductivity of dynamic layer and ease availability of the charge transport through that dynamical layer. Thus, a PCE of 1.28% is attained. Further, aniline functionalized GQDs act as an electron receiver in P3HT surface on optimized cell structure that led to improved performance.[44]

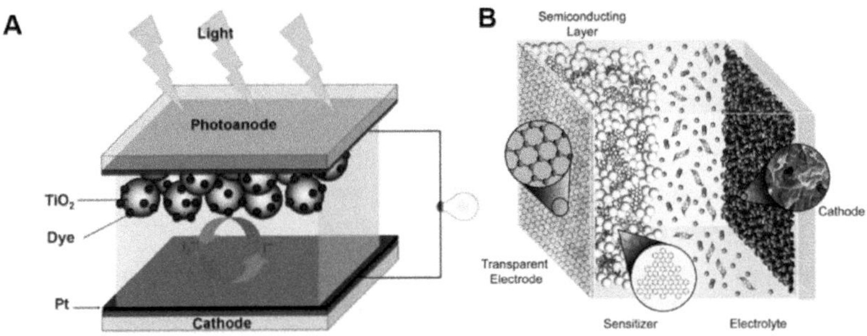

FIGURE 2.20 Schematic representation of (**A**) device structure of DSSC[6] and (**B**) DSSC with graphene used in different components.[80]

Source: Reprinted with permission from a) Ref. [6]. © 2012 Springer Nature; b) © 2014 American Chemical Society.

The edged carboxylated graphene nanoplatelets (ECGnPs) prepared by ball-milling method are employed as an oxygen-rich metal-free CEs. The ECGnPs-based DSSC showed excellent enhancements in electrochemical stability and charge transfer performance for the $Co(bpy)_3^{2+/3+}$ redox couple because of the high charge polarization from variation in electronegativity values among carbon (2.55) and oxygen (3.50). Further, the ECGnPs-based DSSC showed an efficiency of 9.31%.[71]

Basal plane-functionalized GO can be utilized as hole extraction layer as well as electron transport material in active layer in PSCs. The sulfated GO ($GO–OSO_3H$) was synthesized by replacing the in-plane epoxy and/or hydroxyl groups of GO with $–OSO_3H$ groups, however, high retention in edge –COOH groups. The reaction of GO with fuming H_2SO_4 can also eliminate a fraction of the epoxy groups and –OH groups on the basal plane to produce rGO. In comparison with GO, the GO–OSO_3H together with the peripheral –COOH groups, could further result in solubility for solution casting and boost the doping of the donor polymers. In particular, a PSC based on P3HT:PCBM active layer and GO–OSO_3H as hole extraction layer demonstrated a fill factor (FF) of 0.71 and PCE of 4.37%. Both of these values are the maximum recorded for P3HT:PCBM devices. C_{60} grafted graphene can behave as a suitable electron-accepting/transporting materials in PSCs.[83]

Graphene film with a thickness of 10 nm, which is having transparency greater than 70% and conductivity of several hundred Scm^{-1} is obtained by exfoliating graphitic oxide as a starting material followed by thermal reduction. The experimental results encouraged for exploring substitution system for ITO electrode, however, the PCE of the graphene-based solar cell is not efficient as compared with the FTO-based reference device. The PCE of this graphene-based cell is 1.53% that is almost equal to the ITO-based cell (1.5%) in the monochromatic light of wavelength 510 nm. But in the simulated solar radiation, the cell containing graphene and ITO as electrode materials have efficiency of 0.29% and 1.17%, respectively. Generally, Pt sheets on the transparent conducting oxide (TCO) glass are often used as counter electrode material for DSSCs due to their high activity. Though, the price of DSSCs making primarily arises from TCO and dyes, it is essential to substitute TCO by graphene sheet than to substitute Pt by graphene sheet for cost-effectiveness of DSSC.[84] Xu et al. proposed a counter electrode for DSSC, wherein the 1-pyrenebutyrate functionalized graphene is included with FTO (G-FTO) to prepare electrode. The PCE value is 2.2% with G-FTO as counter electrode, which is better than that of the cell containing only FTO as counter electrode (0.048%) but poorer than that of Pt as counter electrode (3.98%).[85] In addition to this, graphene doped with PEDOT:PSS have been used in DSSC. The device containing graphene/PEDOT:PSS as counter electrode exhibited a PCE of 4.5% that is lower than that of cells with Pt as counter electrode (6.3%), also higher than that of cells with PEDOT:PSS as counter electrode. Using solution processed poly(3-octylthiophene) (P3OT) and P3HT as electron donor and functionalized graphene (synthesized by the reaction of exfoliated graphite oxide with phenyl isocyanate) as electron acceptor, a solar cell having PCE 1.4% was resulted under simulated solar radiation at optimized conditions.[84]

Yun and his co-author utilized GO, a conventional rGO, and pr-GO which is a GO being reduced by p-toluenesulphonyl hydrazine (p-TosNHNH$_2$) as the hole transport layer (HTL) for highly efficient solar cells. The cell containing pr-GO showed outstanding function with a power conversion efficiency of 3.63%.[86] Wang et al. assembled an organic photovoltaic (OPV) cell with a graphene electrode, which obtained a PCE of about 3.98%, which is comparatively higher than that of the ITO-based device (3.86%), signifying layer-transferred graphene was assured to be of good quality, affordable, and pliable material for translucency of electrodes in solar cells.[87] Moreover, graphene can also be used as a substituent for counter electrodes for DSSCs, which is because of its exceptional performances, such as huge surface area, good optical transmittance, and outstanding electro catalytic behavior.[88] Kavan et al. set down graphene nanoplatelets (GNP) through thin semi-transparent layer on F-doped SnO$_2$ (FTO), and it results in the PCE values between 8–10%. Furthermore, graphene was also used to create a standard electrode with Pt.[89] Yen et al. prepared a counter electrode by layering Pt NPs/graphene (PtNPs/G) nanocomposite on FTO surface.[90] Their work exhibited the DSSCs included with PtNPs/G nanocomposite with a conversion efficiency of 6.35% that is comparatively 20% more than that of the devices having platinized FTO. It is concluded that the application of graphene intensifies the activity of DSSCs as well as lowers the cost of DSSCs. A comparison of maximum PCE magnitude of solar cells with graphene-based materials and conventional materials as control is presented in Figure 2.21.[88–90]

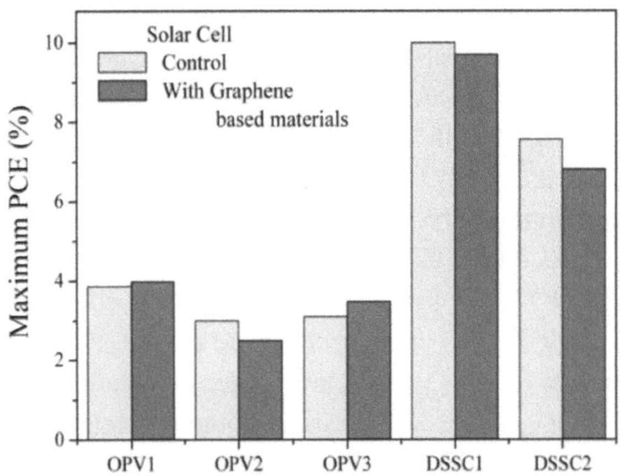

FIGURE 2.21 Comparison of maximum PCE values of solar cell containing graphene-based materials and conventional material.[88]

Source: Reprinted with permission from Ref. [88]. © 2012 John Wiley.

Heteroatom-doped graphene can substitute the high-cost Pt counter electrode in DSSC. Xue et al. verified a counter electrode (CE) formulated on 3D N-doped graphene foams (N-GFs) with a doping range of 7.6% and derived DSSC with PCE of about 7.07% that is equivalent to that of Pt counter electrode (7.44%).[75] Schematic route of the preparation of the N-GF counter electrode from GO suspension is given in Figure 2.22.[91,92]

FIGURE 2.22 Schematic representation of preparation pathway of N-GF counter electrode from GO suspension.[75]

Source: Reprinted with permission from Ref. [75]. © 2012 John Wiley.

Porphyrin-based photovoltaic structures containing Zn porphyrin and ZnO NPs were used as model systems to probe the efficacy of RGO in endorsing charge separation and charge transport in DSSCs. In particular, the ability of RGO to transfer electrons can be employed to encourage charge transport within mesoscopic TiO_2 system. The inorganic/organic structures made from RGO/ZnO/porphyrin uses a special electron transfer cascade opening with photoexcitation of Zn porphyrin. Photoelectrons in Zn porphyrin are transported into the ZnO NPs followed by RGO where they are shifted to SnO_2 layer on the electrode surface. Even though there is involvement of several steps in the electron transfer process, this ternary system demonstrates the effective generation of photocurrents with incident photons to current efficiency (IPCE) of about ~70%. In addition to this, increased photoelectrochemical behavior was also displayed by utilizing the CdSe–RGO nanocomposite as photoanodes in QDSCs.[93]

Particularly, the edge-functionalized graphene materials (EFGs) were widely employed as a novel group of charge transportation and extraction materials in PSCs. Specific EFGs were utilized as extremely effective counter electrodes to substitute Pt in DSSCs caused by the large electrical conductivity and electrocatalytic activity, thus the resulting performance is comparable to conventional Pt-based DSSCs. A DSSC based on nitrogen edge-doped graphene nanoplatelets (NGnPs) cathode exhibited PEC of 9.34% as compared with PEC of 8.85% for Pt-based DSSC. Similarly, DSSCs with edge selectively halogenated (X = Cl, Br, I) *graphene* nanoplatelets (XGnPs), for example, IGnP counter electrode or FGnP counter electrode showed fill factors (FFs) of 71.3% and 71.5%, and PCE values of 10.31% and 10.01%, respectively. In addition, edge-carboxylated graphene nanoplatelets (ECGnPs) or NGnPs-based DSSCs also demonstrated FF of 74.4% and 71.9%, and PCE of 9.31% and 10.27%, respectively.[94]

Semi-transparent perovskite solar cells were prepared by laminating stacked CVD graphene as top transparent electrode on the perovskite layers. The device functioning was developed by enhancing the conductivity of graphene electrodes and contact between the top graphene electrodes and the HTL (spiro-OMeTAD) on perovskite layers. The device containing double-layer graphene electrodes exhibits highest PCE of 12.02% and 11.65% from FTO and graphene sides, respectively, these are reasonably more in comparison with the semi-transparent perovskite solar cells.[95]

It is demonstrated that a single-layer graphene/n-Si Schottky junction solar cells, which show a native power conversion efficiency of 1.9% under sun AM1.5G illumination that upon chemical charge transfer doping with the bis(trifluoromethanesulfonyl)-amide [(($CF_3SO_2)_2NH$)] (TFSA) enhance the device' PCE value to 8.6%. The TFSA dopant gives environmental stability due to its hydrophobic nature. The graphene-based Schottky junction solar cells are advantageous in comparison with ITO Si junctions. The schematic graphene/n-Si (Fig. 2.23a), geometry of TFSA-doped graphene/n-Si Schottky solar cell (Fig. 2.23b), and the optical image of a completed TFSA-doped graphene/n-Si solar cell showing contacts and contact leads (Fig. 2.23c) are represented in Figure 2.23.[96]

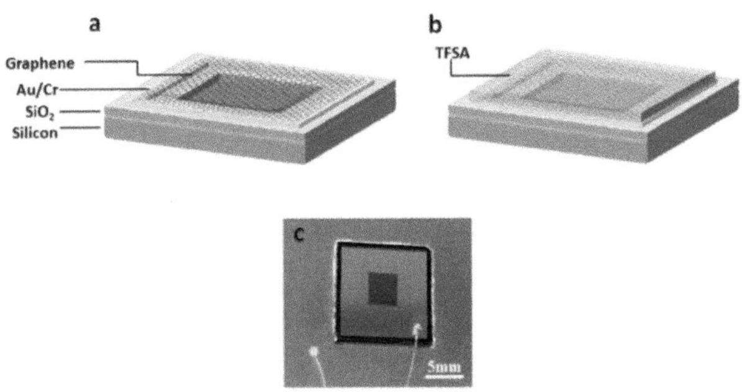

FIGURE 2.23 Schematic representation of graphene/n-Si in solar cell.[96]

Source: Reprinted with permission from Ref. [96]. © 2012 American Chemical Society.

2.5.2 FUEL CELL

Fuel cells are regarded as the inexpensive and eco-friendly conversion technique. The fuel cells transform chemical energy straightforwardly into electrical energy as an alternative of conventional fuel to produce heat energy. The electrochemical cell consists of anode, electrolyte, and cathode materials. For example, when H_2 gas is passed to the anodic electrode, the H_2 is splitting into its constituent electrons and protons. Then the protons drive through the cell toward cathodic electrode while the electrons move out of anode to supply electrical power. Both the electrons and protons finished up at cathode and thus united with O_2 gas to generate H_2O and heat.[6,51] Schematic representation of working principle of fuel cell is presented in Figure 2.24a. The proton exchange membrane in this fuel cell is represented in Figure 2.24b.[97]

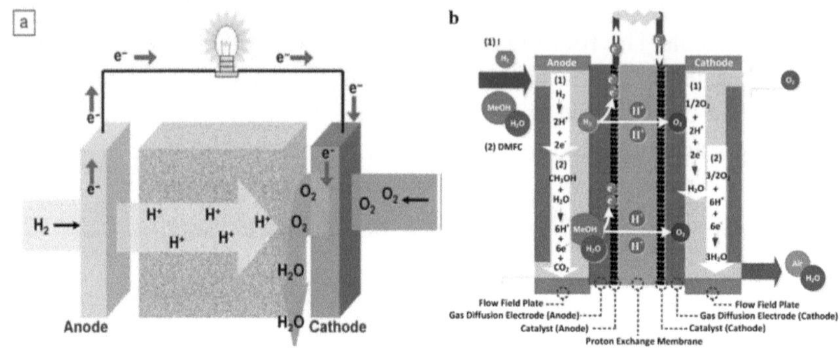

FIGURE 2.24 Schematic representation of (a) working principle of fuel cell[6] and (b) proton exchange membrane in fuel cell.[97]

Source: Reprinted with permission from a) Ref. [6]. © 2012 Springer Nature; b) Ref. [97]. © MDPI open access.

GRMs are an ideal material to smooth the transportation of electrons generated through both fuel oxidation and oxygen reduction reaction (ORR). Further, GRMs were recognized as a promising candidate because proton membranes owing to their high proton conductivity. Various GRMs are proposed both as replacement to metal catalysts or to be employed in conjugation with Pt in composite structures. Pt impregnated graphene and Pt–Ru NPs have high CH_3OH and C_2H_5OH oxidations as compared with the usually employed Vulcan XC-72R carbon black catalyst. RGO alters the electrocatalyst characteristics of Pt infused on its surface. Pt/graphene composite-based electrocatalyst exhibited high activity for CH_3OH oxidation in comparison with the conventional Pt/C black. GRMs lateral size and thickness can be tuned, therefore improving their edge to bulk atoms ratio that can enhance the catalytic activity due to large amount of active catalytic sites present at edges for fuel oxidation at anode and oxygen reduction at cathode. Thus, making GRMs as a potential candidate for the fabrication of cheap and useful fuel cells.[98]

In fuel cells, graphene exhibits as either an active inherent catalyst or a support for catalysts. Graphene has qualified as a multifaceted field to represent fuel cell catalysts because of its chemical stability, conductivity, and high surface area. The developing functionalization of graphene and graphene-based materials with their basic catalytic application makes metal catalysts noble. The nitrogen-doped (N-doped) graphene is the most abundant metal to support fuel cell technologies. The digital image of a transparent N-graphene sheet floating on the water surface is presented in Figure 2.25a. The rotating disk electrode voltammograms for the ORR in air saturated 0.1 M KOH at C-graphene electrode, Pt/C electrode, and N-graphene electrode at electrode rotating rate of 1000 rpm and scan rate of 0.01 V/s are presented in Figure 2.25b.[91]

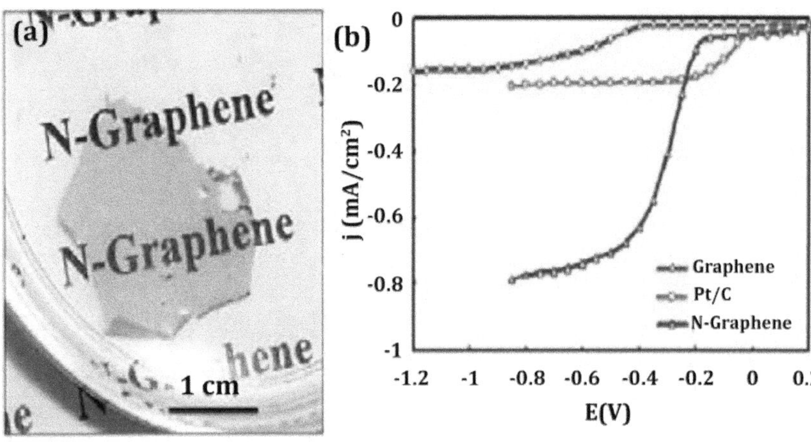

FIGURE 2.25 Schematic representation of N-doped graphene and its electrochemical activity.[6,91]

Source: Reprinted with permission from Ref. [6]. © 2012 Springer Nature.

Qu et al. described the comparison of ORR of graphene and N-graphene synthesized through CVD toward oxygen reduction reaction (ORR) with conventional Pt-filled carbon. It was demonstrated that the N-graphene functions as a metal-free electrode with a significantly improved electro-catalytic activity, high operational stability, and tolerance to crossover effect than Pt for ORR through a four-electron route in alkaline medium resulting in H_2O.[99] Yoo and co-authors synthesized Pt catalysts bearing GNS from a mixture of Pt precursor [Pt $(NO_2)_2.(NH_3)_2$] and GNS diffused in C_2H_5OH. The high resolution TEM (HRTEM) image of Pt/GNS is represented in Figure 2.26A.[9,100] Xin et al. have reported that the utilization of a Pt/GNS catalyst showed a high catalytic activity for both CH_3OH oxidation and ORR in comparison with Pt/C. The CV diagrams are presented in Figure 2.26B. The peak current toward CH_3OH oxidation is displayed and apparently, the current density of Pt/GNS catalyst (182.6 $mAmg^{-1}$) betters the result of Pt/C (77.9 $mAmg^{-1}$). Nevertheless, it is notable that the prior heat treatment of Pt/GNS catalyst enhanced the performance by ~3.5 times over that of Pt/C response. Furthermore, it was described that all of the catalysts deteriorated quickly in the preliminary periods and slowly reached a steady state. It was clear that the highest activity and stability was attained by the heat-treated Pt/GNS followed by Pt/GNS and Pt/C, respectively.[9,54,101]

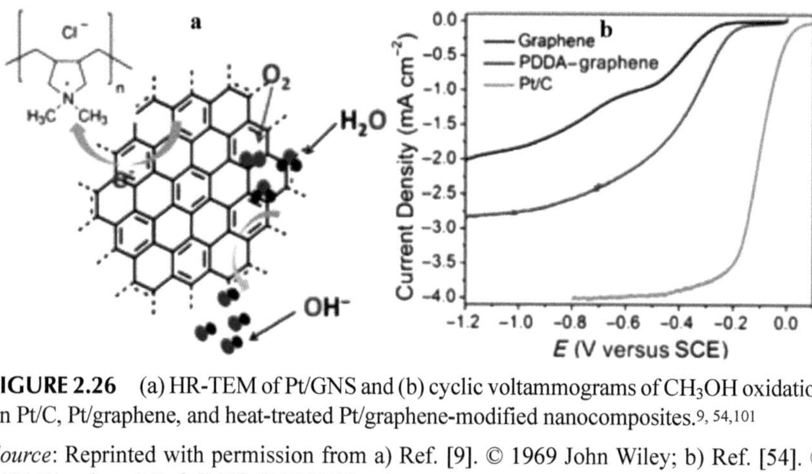

FIGURE 2.26 (a) HR-TEM of Pt/GNS and (b) cyclic voltammograms of CH_3OH oxidation on Pt/C, Pt/graphene, and heat-treated Pt/graphene-modified nanocomposites.[9, 54,101]

Source: Reprinted with permission from a) Ref. [9]. © 1969 John Wiley; b) Ref. [54]. © 2011 Elsevier; c) Ref. [101]. © 2011 Elsevier.

Gong et al. reported that vertically aligned nitrogen-doped carbon nanotubes (VA-CNTs) can successfully catalyze a four-electron ORR process with higher catalytic activity, smaller crossover sensitivity, lower over potential, and good long-term operational stability than commercially available Pt/C black catalysts. Additionally, graphene functionalized with polydiallyl dimethylammonium chloride (PDDA) was demonstrated to display valuable ORR electrocatalytic activity caused by charge transfer between N-free graphene and PDDA. The schematic representation of ORR catalyst based on PDDA noncovalently functionalized graphene is shown in Figure 2.27a. The voltammograms of all ORR on a PDDA/graphene electrode in O_2-saturated 0.1M KOH solution is presented in Figure 2.27b.[6,102]

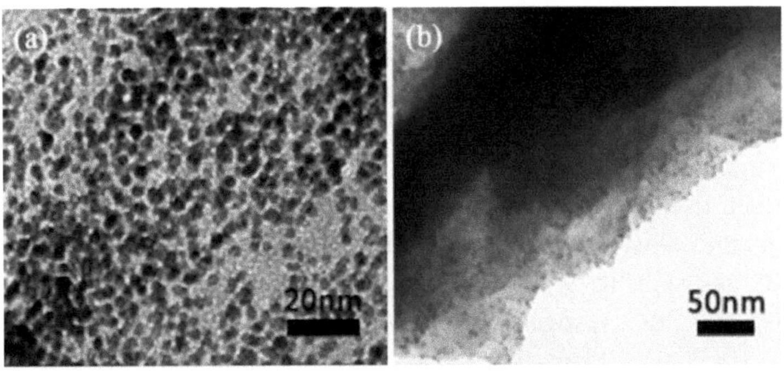

FIGURE 2.27 (a) ORR catalytic activity of PDDA functionalized graphene and (b) voltammograms of all ORR.[6]

Source: Reprinted with permission from Ref. [6]. © 2012 Springer Nature.

It is both theoretically and experimentally adopted that the inclusion of heteroatoms like N, P, and B into carbon having sp^2 hybridization in graphene is mostly helpful to enhance their electrical and chemical activities. Elements having similar electronegativity to carbon (2.55), such as S (2.58) and Se (2.55) were being doped within graphene and also utilized as cathode for ORR. The outcome displayed that the electrocatalytic activities of the Se/S-doped graphene showed remarkable catalytic property, good stability, and high CH_3OH forbearance in alkaline medium for the ORR.[10] Yang and co-author synthesized S-doped graphene, and it has been shown that the electronegativity of doping elements is not significant, however, an essential part is the disruption of electrical neutrality of graphitic planes, and that improves the adsorption of O_2 and becomes the center for the catalysis of O_2, thereby ensuring in magnified activity of doped graphene.[103] Mahmood et al. synthesized graphene-based cobalt sulfide nanocomposite and proposed that the presence of graphene increases the conductivity, has come up with a substantial surface area and the activated centers for additional O_2 adsorption, thus there was an increase in the activity of ORR. This nanocomposite also showed catalytic effect for the ORR in acidic media.[104] Zhang et al. synthesized a N-graphene/iron phthalocyanine (FePc) nanocomposite, a current type of metal electrocatalysts. They displayed that the combination of N-graphene with FePc has come up with a multiple-active catalytic centers to produce an electrically active catalyst which has a strong activity toward catalysis with better conductivity, ensuring a remarkable activity, and a top resistance to fuel than commercial Pt/C.[105] In another work, Zhang et al. reported that the P-doped graphene could improve the ORR on cathode of fuel cells superior than ordinary graphene. This improvement is because of phosphorus activation on neutral π-electronic hindrance of graphene.[106] Wu et al. described a pathway to enhance the catalytic activity of catalysts by constructing a 3D network structure of the graphene that comes up with a bigger surface area and multidimensional layer for the flow of O_2 than the conventional 2D GS.[107]

Pliable paper of graphene with corresponding electrically deposited Cu nanocubes was utilized as anodes in the fuel cells of hydrazine. A paper of carbon fiber overlay with graphene coating can also be utilized as the pliable backing of catalysts in anodic surface. Porous Pd NP catalyst was electrodeposited on the rGO/CF paper that showed high catalytic activity and stability toward C_2H_5OH oxidation. On the other hand, graphene materials were also used as catalyst supports of cathodes in flexible fuel cells. For instance, a N-doped rGO/CNT/Co_3O_4 paper was fabricated for ORR. In addition to the support of catalyst, the inclusion of N into the graphene can access catalytic sites for ORR. The CNTs were organized within the GNSs like sandwiched form to avoid reassembling of GS and enhance the electrical conductivity power of the cathode. The pliable catalyst paper exhibited remarkable longevity and forbearance to poisoning effect of methanol to that of normal Pt/C catalyst. Freshly, the graphene-based compounds have also been discovered for pliable microbial fuel cells. To generate electricity, bacteria within those microbial fuel cells decomposed the organic wastes within water. For more pliability Mink et al. proposed a microbial fuel cell having micrometer in size, which contains anode of graphene, a cathode, and a rubber substrate.[33,108]

It is reported that the polymer/graphene nanocomposites are promising catalyst in fuel cells due to their outstanding electrical conductivity, extraordinary mechanical properties, adequate surface functionality, and excellent electrochemical properties.[37] Gnana et al. have fabricated rGO with modification of PPy through in situ polymerization and bioreduction method. This rGO/PPy showed a high electrocatalytic oxidation current and a high microbial fuel cell (MFC) power density value of 1068 mWm^{-2} including outstanding durability performance of 300 h.[109] Additionally, metals can also be incorporated into the polymer/graphene nanocomposite to enhance the efficiency of fuel cells. Yang et al. described the deposition of Pd NPs on the rGO/PPy nanocomposite for the electrooxidation of formic acid.[110] The Pd NPs are uniformly distributed on the surface of the rGO/PPy nanocomposite as shown in Figure 2.28a. Further, the introduction of Pt NPs into the PANI/rGO nanocomposites to enhance their performance as anode catalysts in microbial fuel cells (MFCs) is presented in Figure 2.28b. A strong covalent bond among PANI and Pt proficiently restricts the accumulation of Pt NPs on the rGO surface thereby resulting in a uniform dispersion of Pt NPs on the graphene surface, which improves the electrical conductivity of the rGO/PANI/Pt. In addition, the rGO/PANI/Pt ternary nanocomposite exhibits high MFC power density and long-life durability.[37]

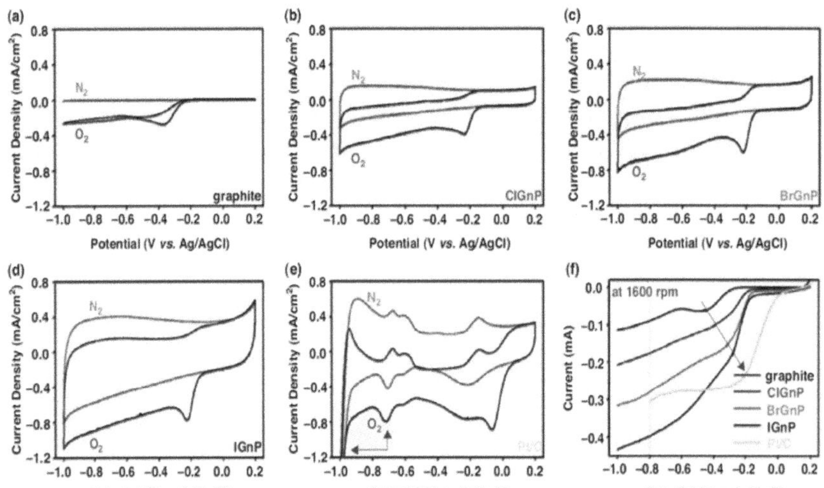

FIGURE 2.28 TEM images of (**a**) Pd/rGO/PPy and (**b**) rGO/PANI/Pt nanocomposites.[37]
Source: Reprinted with permission from Ref. [37]. © 2016 John Wiley.

Simple VAGNAs and Pt NPs reinforced VAGNAs were introduced as electrodes for ORR. The N-doped VAGNAs were shown to enhance the kinetic current and the selectivity for ORR. The Pt/VAGNA electrode displays better catalytic activity than the commercial Pt/C black. The Pt NPs/VAGNA electrodes were also illustrated to show maximum current density for the oxidation of methanol comparatively to that of Pt/C black electrodes, and it has been observed that the electrodes have more resistant to that of CO poisoning. In comparisons with vertically oriented carbon nanofiber, the Pt NPs/VAGNAs could show more efficiency toward longevity and electrochemical activity due to the vulnerability of an excess quantity of strongly reactive plane edges and the vertically interconnected GNSs.[42] Although, the Pt-based catalysts were considered as a suitable catalyst for ORR, the excessive cost and low abundance of novel metal catalysts required for catalyzing the ORR in fuel cells have strongly restricted the fabrication of fuel cell systems. The simple ball milling method has implemented for industrial scale production of the edge-doped GS at affordable price, which is a potential metal-free ORR catalysts with an outstanding durability and hindrance to the CH_3OH crossover/ CO poisoning effects. The ball-milling method was employed to fabricate not only N-doped EFGs but also different heteroatoms-doped EFGs for the ORR, which includes hydrogen (HGnP), sulfonic acid groups (SGnP), carboxylic acid/sulfonic acid groups (CSGnP) GnPs. The relative order of the ORR electrocatalytic activity of these EFGs are in the decreasing order of SGnP>CSGnP>CGnP>HGnP>pristine graphite, which informed that both the oxygen diffusion kinetics and edge polarity of the heteroatom-doped EFGs can considerably promote the ORR. Similarly, in the case of XGnPs, the observed electrocatalytic activity for ORR are in decreasing order of IGnP>BrGnP>ClGnP.[94]

The incorporation of heteroatoms in the GS can modify the electronic structure of the graphene network for improving the catalytic activity because of the electronegativity differences among doped heteroatoms and carbon atoms. The adsorption of oxygen molecules and charge transport behavior was enhanced and thereby increasing the ORR activity due to the polarization of adjacent carbon atoms by heteroatoms in graphene network. Therefore, all the resultant functionalized GS exhibited significant ORR activities, which is improved than the conventional Pt catalyst.

Between all the XGnP electrodes, the IGnP presented the highest capacitances of 127.6 and 139.5 F/g both in N_2 and O_2 saturated electrolytes with a high cycle stability. The cyclic voltammograms (CV) of (a) pristine graphite, (b) ClGnP, (c) BrGnP, (d) IGnP, and (e) Pt/C on glassy carbon electrode in N_2 and O_2 saturated 0.1 M KOH aqueous solution at the scan rate of 10 mV/s and (f) linear sweep voltammograms (LSV) are represented in Figure 2.29.[71]

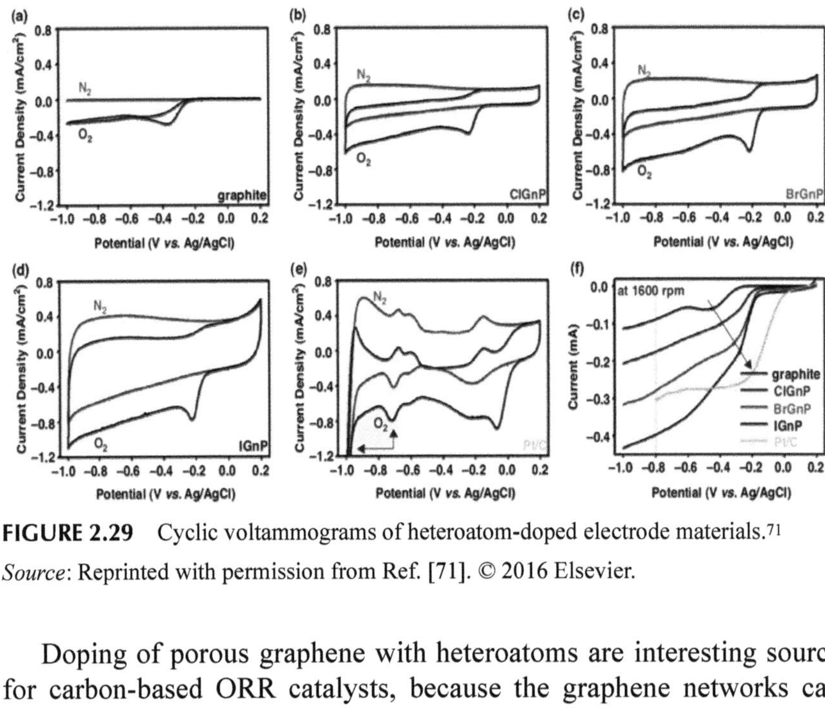

FIGURE 2.29 Cyclic voltammograms of heteroatom-doped electrode materials.[71]
Source: Reprinted with permission from Ref. [71]. © 2016 Elsevier.

Doping of porous graphene with heteroatoms are interesting source for carbon-based ORR catalysts, because the graphene networks can be favorable for the transport of electron. The electroactivity of 2D graphene-based carbon nitride (G-CN) nanosheets for the ORR in an alkaline electrolytic solution was being observed with CV and rotating disk electrode voltammetry. By the process of pyrolysis, G-CN nanosheets at a temperature of 800°C showed a maximal kinetic current density (involvement of four-electron transfer mechanism), that is higher than that of the commercially available Pt/C electrode. To fabricate high-efficient ORR catalyst, another strategy is applied which is the incorporation of oxides of metal or non-valuable metal within nitrogen-doped porous graphene networks. In addition, a N-doped graphene aerogel

(N-GA) packed with Fe_3O_4 NPs (Fe_3O_4/N-GAs) was produced and showed a maximal BET surface area and pronounced electrocatalytic ORR activity in KOH electrolyte.[92]

2.6 ENVIRONMENTAL EFFECT OF THE GRAPHENE

Graphene and graphene-based materials can also be utilized for environmental purposes. Applications of graphene-based materials in environmental aspects can be summarized into two categories, which are environmental protection and detection. The functionalized graphene for environmental protection includes adsorption and reduction of heavy metal ions and adsorption of organic pollutants. On the other hand, functionalized graphene for environmental detection includes the detection of toxic gases in air, heavy metals ions, and organic pollutants.[111]

2.7 CONCLUSIONS

In this review, the structure, properties, prehistory, method of preparations, applications of graphene by manipulating its properties, that is, by functionalization of graphene are described. Furthermore, the applications of graphene and graphene-based materials in energy storage, especially supercapacitor, lithium-ion battery and energy-conversion systems particularly solar cell, fuel cell are well summarized in this review. The charge storage capacity of supercapacitors and Li-ion range batteries are enhanced by the implementation of graphene-based materials in their structural parts. Similarly, the energy conversion in solar cells and fuel cells are also improved by applying graphene-based materials.

KEYWORDS

- **graphene**
- **supercapacitor**
- **lithium-ion battery**
- **solar cell**
- **fuel cell**

REFERENCES

1. Pumera, M. *Energ. Environ. Sci.* **2011,** *4* (3), 668–674.
2. Zhu, J.; Yang, D.; Yin, Z.; Yan, Q.; Zhang, H. *Small* **2014,** *10* (17), 3480–3498.
3. Coroş, M.; Pogăcean, F.; Măgeruşan, L.; Socaci, C.; Pruneanu, S. *Front. Mater. Sci.* **2019,** *13* (1), 23–32.
4. Berry, V. *Carbon* **2013,** *62*, 1–10.
5. Geim, A. K.; Novoselov, K. S. *Nat. Mater.* **2007,** *6* (3), 183–191.
6. Liu, J.; Xue, Y.; Zhang, M.; Dai, L. *MRS Bull.* **2012,** *37* (12), 1265–1272.
7. Hou, J.; Shao, Y.; Ellis, M. W.; Moore, R. B.; Yi, B. *Phys. Chem. Chem. Phys.* **2011,** *13* (34), 15384–15402.
8. Ye, M.; Zhang, Z.; Zhao, Y.; Qu, L. *Joule* **2018,** *2* (2), 245–268..
9. Luo, B.; Liu, S.; Zhi, L. *Small* **2011,** *8* (5), 630–646.
10. Mahmood, N.; Zhang, C.; Yin, H.; Hou, Y. *J. Mater. Chem. A* **2014,** *2* (1), 15–32.
11. Wang, H.; Feng, H.; Li, J. *Small* **2014,** *10* (11), 2165–2181.
12. Geim, A. K. *Phys. Scr.* **2012,** T146 014003 (4pp).
13. Brodie B. C. *Phil. Trans. R. Soc.* **1859,** *149*, 249–259.
14. Ruess, G.; Vogt, F. *Monatshefte Chem.* **1948,** *78*, 222–242.
15. Boehm, H. P.; Clauss, A.; Fischer, G. O.; Hofmann, U. *Z. Anorg. Allg. Chem.* **1962,** *316*, 119–127.
16. Boehm, H. P.; Setton, R.; Stumpp, E. *Carbon* **1986,** *24* (2), 241–245.
17. Novoselov, K. S.; Geim, A. K.; Morozov, S. V.; Jiang, D.; Zhang, Y.; Dubonos, S. V.; Grigorieva, I. V. Firsov, A. A. *Science* **2004,** *306* (5696), 666–669.
18. Allen, M. J.; Tung, V. C.; Kaner, R. B. *Chem. Rev.* **2010,** *110* (1), 132–145.
19. Yang, Y.; Han, C.; Jiang, B.; Iocozzia, J.; He, C.; Shi, D.; Jiang, T.; Lin, Z. *Mater. Sci. Eng. R: Rep.* **2016,** *102*, 1–72.
20. Raccichini, R.; Varzi, A.; Passerini, S.; Scrosati, B. *Nat. Mater.* **2014,** *14* (3), 271–279.
21. Novoselov, K. S.; Falko, V. I.; Colombo, L.; Gellert, P. R.; Schwab, M. G.; Kim, K. *Nature* **2012,** *490* (7419), 192–200.
22. Zhang, Q.; Uchaker, E.; Candelaria, S. L.; Cao, G. *Chem. Soc. Rev.* **2013,** *42* (7), 3127–3171.
23. Choi, H.-J.; Jung, S.-M.; Seo, J.-M.; Chang, D. W.; Dai, L.; Baek, J.-B. *Nano Energ.* **2012,** *1* (4), 534–551.
24. Chen, X.; Paul, R.; Dai, L. *Natl. Sci. Rev.***2017,** *4* (3), 453–489.
25. Stoller, M. D.; Park, S.; Zhu, Y.; An, J.; Ruoff, R. S. *Nano Lett.* **2008,** *8* (10), 3498–3502.
26. Wang, Y.; Shi, Z.; Huang, Y.; Ma, Y.; Wang, C.; Chen, M.; Chen, Y. *J. Phys. Chem. C* **2009,** *113* (30), 13103–13107.
27. Du, Q.; Zheng, M.; Zhang, L.; Wang, Y.; Chen, J.; Xue, L.; Dai, W.; Ji, G.; Cao, J. *Electrochim. Acta* **2010,** *55* (12), 3897–3903.
28. Lv, W.; Tang, D.-M.; He, Y.-B.; You, C.-H.; Shi, Z.-Q.; Chen, X.-C.; Chen, C.-M.; Hou, P.-X.; Liu, C.; Yang, Q.-H. *ACS Nano* **2009,** *3* (11), 3730–3736.
29. Yang, X.; Zhu, J.; Qiu, L.; Li, D. *Adv. Mater.* **2011,** *23* (25), 2833–2838.

30. Wang, D.-W.; Li, F.; Zhao, J.; Ren, W.; Chen, Z.-G.; Tan, J.; Wu, Z.-S.; Gentle, I.; Lu, G. Q. Cheng, H.-M. *ACS Nano* **2009,** *3* (7), 1745–1752.
31. Du, F.; Yu, D.; Dai, L.; Ganguli, S.; Varshney, V.; Roy, A. K. *Chem. Mater.* **2011,** *23* (21), 4810–4816.
32. Liu, C.; Yu, Z.; Neff, D.; Zhamu, A.; Jang, B. Z. *Nano Lett.* **2010,** *10* (12), 4863–4868.
33. Wang, X.; Shi, G. *Energ. Environ. Sci.* **2015,** *8* (3), 790–823.
34. Mao, J.; Iocozzia, J.; Huang, J.; Meng, K.; Lai, Y.; Lin, Z. *Energ. Environ. Sci.* **2018,** *11* (4), 772–799.
35. Zhang, L.; Shi, G. *J. Phys. Chem. C* **2011,** *115* (34), 17206–17212.
36. Luan, V. H.; Tien, H. N.; Hoa, L. T.; Hien, N. T. M.; Oh, E.-S.; Chung, J.; Kim, E. J.; Choi, W. M.; Kong, B.-S.; Hur, S. H. *J. Mater. Chem. A* **2013,** *1* (2), 208–211.
37. Cui, L.; Gao, J.; Xu, T.; Zhao, Y.; Qu, L. *Chem. Asian J.* **2016,** *11* (8), 1151–1168.
38. Choi, H.; Ahn, K.-J.; Lee, Y.; Noh, S.; Yoon, H. *Adv. Mater. Interf.* **2015,** *2* (12), 1500117.
39. Xiao, F.; Yang, S.; Zhang, Z.; Liu, H.; Xiao, J.; Wan, L.; Luo, J.; Wang, S.; Liu, Y. *Sci. Rep.***2015,** *5* (1), 9359 (8pp).
40. Cao, X.; Shi, Y.; Shi, W.; Lu, G.; Huang, X.; Yan, Q.; Zhang, Q.; Zhang, H. *Small* **2011,** *7* (22), 3163–3168.
41. Chen, C.-M.; Zhang, Q.; Zhao, X.-C.; Zhang, B.; Kong, Q.-Q.; Yang, M.-G.; Yang, Q.-H.; Wang, M.-Z.; Yang, Y.-G.; Schlögl, R.; Su, D. S. *J. Mater. Chem.* **2012,** *22* (28), 14076–14084.
42. Zhang, Z.; Lee, C.-S.; Zhang, W. *Adv. Energ. Mater.* **2017,** *7* (23), 1700678 (20pp).
43. Wu, Q.; Xu, Y.; Yao, Z.; Liu, A.; Shi, G. *ACS Nano* **2010,** *4* (4), 1963–1970.
44. Hu, C.; Song, L.; Zhang, Z.; Chen, N.; Feng, Z.; Qu, L. *Energ. Environ. Sci.* **2015,** *8* (1), 31–54.
45. Choi, B. G.; Hong, J.; Hong, W. H.; Hammond, P. T.; Park, H. *ACS Nano*, **2011,** *5* (9), 7205–7213.
46. Liu, L.; Niu, Z.; Zhang, L.; Zhou, W.; Chen, X.; Xie, S. *Adv. Mater.* **2014,** *26* (28), 4855–4862.
47. Cong, H.-P.; Ren, X.-C.; Wang, P.; Yu, S.-H. *Energ. Environ. Sci.* **2013,** *6* (4), 1185–1191.
48. Lu, X.; Dou, H.; Yuan, C.; Yang, S.; Hao, L.; Zhang, F.; Shen, L.; Zhang, L.; Zhang, X. *J. Power Sources* **2012,** *197*, 319–324.
49. He, Y. M.; Chen, W. J.; Li, X. D.; Zhang, Z. X.; Fu, J. C.; Zhao, C. H.; Xie, E. Q. *ACS Nano* **2013,** *7* (1), 174–182.
50. Xie, J. L.; Guo, C. X.; Li, C. M. *Energ. Environ. Sci.* **2014,** *7* (8), 2559–2579.
51. Dai, L.; Chang, D. W.; Baek, J.-B.; Lu, W. *Small* **2012,** *8* (8), 1130–1166.
52. Yoo, E.; Kim, J.; Hosono, E.; Zhou, H.; Kudo, T.; Honma, I. *Nano Lett.* **2008,** *8* (8), 2277–2282.
53. Wu, Z.-S.; Ren, W.; Xu, L.; Li, F.; Cheng, H.-M. *ACS Nano* **2011,** *5* (7), 5463–5471.
54. Brownson, D. A. C.; Kampouris, D. K.; Banks, C. E. *J. Power Sources* **2011,** *196* (11), 4873–4885.
55. Paek, S.-M.; Yoo, E.; Honma, I. *Nano Lett.* **2009,** *9* (1), 72–75.

56. Yang, S.; Feng, X.; Wang, L.; Tang, K.; Maier, J.; Müllen, K. *Angew. Chem. Int. Ed.* **2010,** *49* (28), 4795–4799.
57. Fang, Y.; Lv, Y.; Che, R.; Wu, H.; Zhang, X.; Gu, D.; Zheng, G.; Zhao, D. *J. Am. Chem. Soc.* **2013,** *135* (4), 1524–1530.
58. Yin, S.; Zhang, Y.; Kong, J.; Zou, C.; Li, C. M.; Lu, X.; Ma, J.; Boey, F. Y. C. Chen, X. *ACS Nano*, **2011,** *5* (5), 3831–3838.
59. Wang, B.; Al Abdulla, W.; Wang, D.; Zhao, X. S. *Energ. Environ. Sci.* **2015,** *8* (3), 869–875.
60. Wang, Z.-L.; Xu, D.; Xu, J.-J.; Zhang, L.-L.; Zhang, X.-B. *Adv. Funct. Mater.* **2012,** *22* (17), 3699–3705.
61. Hu, G.; Xu, C.; Sun, Z.; Wang, S.; Cheng, H.-M.; Li, F.; Ren, W. *Adv. Mater.* **2015,** *28* (8), 1603–1609.
62. Xiao, L.; Wu, D.; Han, S.; Huang, Y.; Li, S.; He, M.; Zhang, F.; Feng, X. *ACS Appl. Mater. Interf.* **2013,** *5* (9), 3764–3769.
63. Wei, W.; Yang, S.; Zhou, H.; Lieberwirth, I.; Feng, X.; Müllen, K. *Adv. Mater.* **2013,** *25* (21), 2909–2914.
64. Garakani, M. A.; Abouali, S.; Zhang, B.; Takagi, C. A.; Xu, Z.-L.; Huang, J.; Huang, J.; Kim, J.-K. *ACS Appl. Mater. Interf.* **2014,** *6* (21), 18971–18980.
65. Meng, J.; Cao, Y.; Suo, Y.; Liu, Y.; Zhang, J.; Zheng, X. *Electrochim. Acta* **2015,** *176*, 1001–1009.
66. Jeong, S. Y.; Yang, S.; Jeong, S.; Kim, I. J.; Jeong, H. J.; Han, J. T.; Baeg, K.-J.; Lee, G.-W. *Small* **2015,** *11* (23), 2774–2781.
67. Li, M.; Song, H.; Chen, X.; Zhou, J.; Ma, Z. *Phys. Chem. Chem. Phys.* **2015,** *17* (5), 3250–3260.
68. Su, Y.; Liu, Y.; Liu, P.; Wu, D.; Zhuang, X.; Zhang, F.; Feng, X. *Angew. Chem.* **2014,** *127* (6), 1832–1836.
69. Wu, G.; Gao, W. *Graphene Oxide*; Gao, W., Ed.; Springer Int. Pub.: Switzerland, 2015; pp. 97–127.
70. Xia, G.; Li, N.; Li, D.; Liu, R.; Wang, C.; Li, Q.; Lü, X.; Spendelow, J. S.; Zhang, J.; Wu, G. *ACS Appl. Mater. Interf.* **2013,** *5* (17), 8607–8614.
71. Fan, X.; Chang, D. W.; Chen, X.; Baek, J.-B.; Dai, L. *Curr. Opin. Chem. Eng.* **2016,** *11*, 52–58.
72. Chen, D.; Chen, W.; Ma, L.; Ji, G.; Chang, K.; Lee, J. Y. *Mater. Today* **2014,** *17* (4), 184–193.
73. Yu, D.; Yang, Y.; Durstock, M.; Baek, J.-B.; Dai, L. *ACS Nano* **2010,** *4* (10), 5633–5640.
74. Wang, H.; Sun, K.; Tao, F.; Stacchiola, D. J.; Hu, Y. H. *Angew. Chem. Int. Ed.* **2013,** *52* (35), 9210–9214.
75. Xue, Y.; Liu, J.; Chen, H.; Wang, R.; Li, D.; Qu, J.; Dai, L. *Angew. Chem. Int. Ed.* **2012,** *51* (48), 12124–12127.
76. Ahn, H.-J.; Kim, I.-H.; Yoon, J.-C., Kim, S.-I.; Jang, J.-H. *Chem. Comm.* **2014**, *50* (19), 2412–2415.
77. Yuan, S.; Tang, Q.; He, B.; Zhao, Y. *J. Power Sources* **2014,** *260*, 225–232.
78. Kim, H.-N.; Yoo, H.; Moon, J. H. *Nanoscale* **2013,** *5* (10), 4200–4204.

79. Ergen, O.; Gilbert, S. M.; Pham, T.; Turner, S. J.; Tan, M. T. Z.; Worsley, M. A.; Zettl, A. *Nat. Mater.* **2016,** *16* (5), 522–525.
80. Roy-Mayhew, J. D.; Aksay, I. A. *Chem. Rev.***2014,** *114* (12), 6323–6348.
81. Guo, C. X.; Yang, H. B.; Sheng, Z. M.; Lu, Z. S.; Song, Q. L.; Li, C. M. *Angew. Chem. Int. Ed.* **2010,** *49* (17), 3014–3017.
82. Liu, Z. Y.; He, D. W.; Wang, Y. S.; Wu, H. P.; Wang, J. G. *Sol. Energ. Mater. Sol. Cells* **2010,** *94*, 1196–1200.
83. Dai, L. *Acc. Chem. Res.***2012,** *46* (1), 31–42.
84. Liang, M.; Luo, B.; Zhi, L. *Int. J. Energ. Res.* **2009,** *33* (13), 1161–1170.
85. Xu, Y.; Bai, H.; Lu, G.; Li, C.; Shi, G. *J. Am. Chem. Soc.* **2008,** *130* (18), 5856–5857.
86. Yun, J.-M.; Yeo, J.-S.; Kim, J.; Jeong, H.-G.; Kim, D.-Y.; Noh, Y.-J.; Kim, S.-S.; Ku, B.-C.; Na, S.-I. *Adv. Mater.* **2011,** *23* (42), 4923–4928.
87. Wang, Z.; Puls, C. P.; Staley, N. E.; Zhang, Y.; Todd, A.; Xu, J.; Howsare, C. A.; Hollander, M. J.; Robinson, J. A. Liu, Y. *Phys. E: Low Dimens. Syst. Nanostruct.* **2011,** *44* (2), 521–524.
88. Sahoo, N. G.; Pan, Y.; Li, L.; Chan, S. H. *Adv. Mater.***2012,** *24* (30), 4203–4210.
89. Kavan, L.; Yum, J.-H.; Nazeeruddin, M. K.; Gratzel, M. *ACS Nano* **2011,** *5* (11), 9171–9178.
90. Yen, M.-Y.; Teng, C.-C.; Hsiao, M.-C.; Liu, P.-I.; Chuang, W.-P.; Ma, C.-C. M.; Hsieh, C.-K.; Tsai, M.-C.; Tsai, C.-H. *J. Mater. Chem.* **2011,** *21* (34), 12880–12888.
91. Zhang, J.; Zhao, F.; Zhang, Z.; Chen, N.; Qu, L. *Nanoscale* **2013,** *5* (8), 3112–3126.
92. Han, S.; Wu, D.; Li, S.; Zhang, F.; Feng, X. *Adv. Mater.* **2013,** *26* (6), 849–864.
93. Lightcap, I. V.; Kamat, P. V. *Acc. Chem. Res.* **2012,** *46* (10), 2235–2243.
94. Xiang, Z.; Dai, Q.; Chen, J.-F.; Dai, L. *Adv. Mater.* **2016,** *28* (29), 6253–6261.
95. You, P.; Liu, Z.; Tai, Q.; Liu, S.; Yan, F. *Adv. Mater.* **2015,** *27* (24), 3632–3638.
96. Miao, X.; Tongay, S.; Petterson, M. K.; Berke, K.; Rinzler, A. G.; Appleton, B. R.; Hebard, A. F. *Nano Lett.* **2012,** *12* (6), 2745–2750.
97. Ye, Y. S.; Rick, J.; Hwang, B. J. *Polymers*, **2012,** *4* (2), 913–963.
98. Bonaccorso, F.; Colombo, L.; Yu, G.; Stoller, M.; Tozzini, V.; Ferrari, A. C.; Ruoff, R. S.; Pellegrini, V. *Science* **2015,** *347* (6217), 1246501–1246501.
99. Qu, L.; Liu, Y.; Baek, J.-B.; Dai, L. *ACS Nano*, **2010,** *4* (3), 1321–1326.
100. Yoo, E.; Okata, T.; Akita, T.; Kohyama, M.; Nakamura, J.; Honma, I. *Nano Lett.* **2009,** *9* (6), 2255–2259.
101. Xin, Y.; Liu, J.-G.; Zhou, Y.; Liu, W.; Gao, J.; Xie, Y.; Yin, Y.; Zou, Z. *J. Power Sources* **2011,** *196* (3), 1012–1018.
102. Gong, K.; Du, F.; Xia, Z.; Durstock, M.; Dai, L. *Science* **2009,** *323* (5915), 760–764.
103. Yang, Z.; Yao, Z.; Li, G.; Fang, G.; Nie, H.; Liu, Z.; Zhou, X.; Chen, X.; Huang, S. *ACS Nano* **2011,** *6* (1), 205–211.
104. Mahmood, N.; Zhang, C.; Jiang, J.; Liu, F.; Hou, Y. *Chem. Eur. J.* **2013,** *19* (16), 5183–5190.
105. Zhang, C.; Hao, R.; Yin, H.; Liu, F.; Hou, Y. *Nanoscale*, **2012,** *4* (23), 7326–7329.
106. Zhang, C.; Mahmood, N.; Yin, H.; Liu, F.; Hou, Y. *Adv. Mater.* **2013,** *25* (35), 4932–4937.

107. Wu, Z.-S.; Yang, S.; Sun, Y.; Parvez, K.; Feng, X.; Mullen, K. *J. Am. Chem. Soc.* **2012,** *134* (22), 9082–9085.
108. Mink, J. E.; Qaisi, R. M.; Hussain, M. M. *Energ. Technol.* **2013,** *1* (11), 648–652.
109. Gnana kumar, G.; Kirubaharan, C. J.; Udhayakumar, S.; Ramachandran, K.; Karthikeyan, C.; Renganathan, R.; Nahm, K. S. *ACS Sust. Chem. Eng.* **2014,** *2* (10), 2283–2290.
110. Yang, S.; Shen, C.; Liang, Y.; Tong, H.; He, W.; Shi, X.; Zhang, X.; Gao, H. *Nanoscale* **2011,** *3* (8), 3277–3284.
111. Lü, M.; Li, J.; Yang, X.; Zhang, C.; Yang, J.; Hu, H.; Wang, X. *Chin. Sci. Bull.* **2013,** *58* (22), 2698–2710.

CHAPTER 3

CARBON NANOTUBE-FERRITE HYBRID FRAMEWORKS FOR ELECTROMAGNETIC WAVE ATTENUATION AND OTHER POTENTIAL APPLICATIONS

ANN V SONY, AMALA ROSE AUGUSTINE, SNITHA VINOD K, and ANN ROSE ABRAHAM*

Department of Physics, Sacred Heart College (Autonomous), Thevara, Kochi, Kerala 682013, India

**Corresponding author. E-mail: annroseabraham@shcollege.ac.in*

ABSTRACT

In this chapter, we describe the development of carbon nanotube (CNT)–ferrite hybrid frameworks, their characterization techniques and varied applications, particularly electromagnetic interference (EMI) shielding and other potential applications. CNT is an ideal catalyst and ferrites are ferromagnetic materials having innumerable applications in various fields owing to their unique electric and magnetic properties. The discovery of one-dimensional carbon nanotubes (CNTs) by Iijima[30] and the development of carbon nanotube composites was a revolutionary creation. The synthesis of the ferrite–CNT nanocomposites by coprecipitation and other synthesis routes, with special emphasis on the hydrothermal method of synthesis, is discussed in this chapter. The incorporation of CNTs in the

Nanostructured Carbon for Energy Generation, Storage, and Conversion. V. I. Kodolov, DSc, Omari Mukbaniani, DSc, Ann Rose Abraham, PhD, and A. K. Haghi, PhD (Eds.)

nanocomposites has proved to substantially improve the properties of the material such as tensile strength, Young's modulus to a great extent. The ferrite–CNT nanocomposites are an attractive topic of research these days due to its high-end applications. The applications briefed in this review include wastewater treatment, microwave absorption, electrochemical applications, magnetic data storage applications, and photocatalysis.

3.1 INTRODUCTION

Decoration of carbon nanotubes (CNTs) with spinel ferrites offers amazing potential for electromagnetic shielding and electronic applications of the scientific world today. Magnetic nanostructures find innumerable applications in different sectors such as memory devices, medicine, nanotechnology, and optical transducers. Carbon nanotubes (CNTs) have an astounding array of physiochemical properties. Laser ablation,[90] chemical vapor deposition,[70,75] template methods,[41] flame synthesis,[71,94,95] pyrolysis,[10,39] electrolysis,[89] solar approaches,[42] and electron or ion beam irradiation[9,107] are different laboratory and industrial methods for the synthesis of CNTs. The cup-stack CNTs are newly created semiconductor material that contains graphene layers in stacking structure. CNT balls are made by the emulsion method that shows good supercapacitor performance in comparison with CNT films.[94] CNTs have great capabilities to act as substitutes for conventional catalysts. The activity of the catalyst depends on size distributions of active phases and size of the particle. The catalytic selectivity is determined by the nature of support. For some hydrogenation reactions, CNTs and functionalized CNTs act as nonmetallic heterogeneous catalysts. In decalin dehydrogenation assisted with microwave, CNT-supported platinum has a high yield to H_2 in comparison with activated carbon, carbon black, graphite, and carbon nanofibers. In the case of oxidation reactions, CNT-based catalysts possess better conversion than others. These catalysts can be used as an alternative conventional iron-based catalyst in the synthesis of ammonia. It is not easy to control the location and amount of magnetic material inside the tube. The phenomenon of sudden penetration of fluids into the wettable capillaries is the basic idea behind the filling of nanotubes with molten metal. But this method is time-consuming and inefficient for large-scale production and some melts react with carbon also. To avoid these drawbacks, synthesis of CNTs is done by noncatalytic chemical vapor deposition (CVD) within the pores of alumina template; filling is done by suspensions of functional nanoparticles and finally, separated from the alumina membrane. In this way, we can use commercially available ferrofluids for filling CNTs and then by evaporation to produce magnetic nanotubes.

Ferrites are ceramic magnetic materials exhibiting ferromagnetic properties below the Curie temperature (T_c). They are made of iron oxide (Fe_2O_3) and metal oxides of divalent metal ions (Cu, Ni, Mn, Co, and Al). However, the Fe^{3+} ions can be easily substituted by a mixture of other ions. Ferrites can be broadly classified into soft and hard ferrites based on their properties. Their important electrical and magnetic properties include high electrical resistance, high magnetic permeability, dielectric losses, high saturation magnetization, high electrical resistivity, and low eddy currents. Ferrites have a DC resistivity 10^4 to 10^{11} times more than that of iron. The method of preparation, state of synthesis, and the substitution of ions influence their crystallography, electrical and magnetic properties to a great extent. The different types of ferrites include Spinel, Hexagonal, and Garnet ferrites, out of which the spinel and magnetoplumbite hexagonal ferrites are the most widely used since their electromagnetic properties and microwave absorption performance can be easily modified.[91] It is from the two basic metal oxide families—spinels and garnets, all ferrites except hexagonal are derived. The structure of spinel is cubic and contains two sublattices (A and B). But garnets feature three sublattices (a, d, and c). The microwave device depends on the interaction of magnetization of ferrite and magnetic vector of the incident wave. The same device can be made as a microwave switch by reversing the magnetic field's direction. Variation of magnetization with temperature, power losses, shape and steady of hysteresis loop are three different principal areas of choice of ferrite for microwave applications.[14] Ferrites find use in vast areas such as linear, digital, and microwave applications owing to their unique electrical and magnetic properties. Since they possess spontaneous magnetic moment below T_c, they find applications in electrical communication equipment, magnetic memories, microwave absorbers, magnetic recording devices, and memory core in computers.[91]

Lately, researchers have been exploring the possibilities of using ferrite-based photoelectrodes for solar water splitting in the PEC cell for producing clean and storable hydrogen energy. But this is a developing technology and there are a large number of challenges that this technology faces. The most crucial part here is choosing the right light absorber that can transfer solar energy to charge carriers to initiate the water-splitting reaction. Ferrites have small band gaps and proper band positions that are appropriate for absorbing sufficient energy and initiating redox reactions of water splitting. The energy required to perform the water splitting and the additional energy to overcome the kinetic barrier of the reaction are derived from the solar energy through a photocatalytic or photovoltaic light absorber, and the performance of the light absorber primarily determines the solar-to-hydrogen (STH) conversion efficiencies. So, the durability and material of the light absorber play a very important role in the PEC system. The PEC cell is powered by solar energy harvested by a photoelectrode. For the fabrication of an efficient photoelectrode, selecting an effective synthesis method that suits the specific purpose is very important as it determines the intrinsic semiconducting properties. In a photoelectrode, the light absorber and the substrate take up the majority volume and cost of the device. So, using a ferrite-based light absorber would substantially reduce the overall cost of the device. Furthermore, ferrites have superior photostability. Nonetheless, they have indirect bandgap characteristics, poor optoelectronic properties, and poor charge carrier dynamics which is why they do not absorb photons as much as their bandgap energy would allow. Nevertheless, researchers are working on understanding the ferrite materials and improving their efficiency to make this an economically viable and feasible way to produce and store hydrogen energy for various applications.[37]

The development of nanocomposite technology is a rapidly growing field. Nanocomposites incorporating carbon nanotubes (CNTs) or minerals and other nanoparticles or metals have been a research area of high scientific impact. The incorporation of metal nanoparticles contributes to improvement in properties such as Young's modulus, tensile strength, electrical and thermal conductivity, impact and scratch resistance, thermal stability, and fire resistance. In this chapter, we discuss some of the potential applications of ferrite–CNT nanocomposites, which are illustrated in Figure 3.1.

3.2 DEVELOPMENT OF CNT–FERRITE NANOCOMPOSITES

There are many methods for the development of ferrite–CNT nanocomposites such as citrate sol–gel combustion method,[16] microemulsion method,[76] reverse micelle method,[80] self-propagating combustion method, coprecipitation, and solution mixing techniques (Bibi et al., 2017). The hydrothermal method is a significant method for synthesis of carbon nanotubes/ ferrite hybrid nanocomposites.[34]

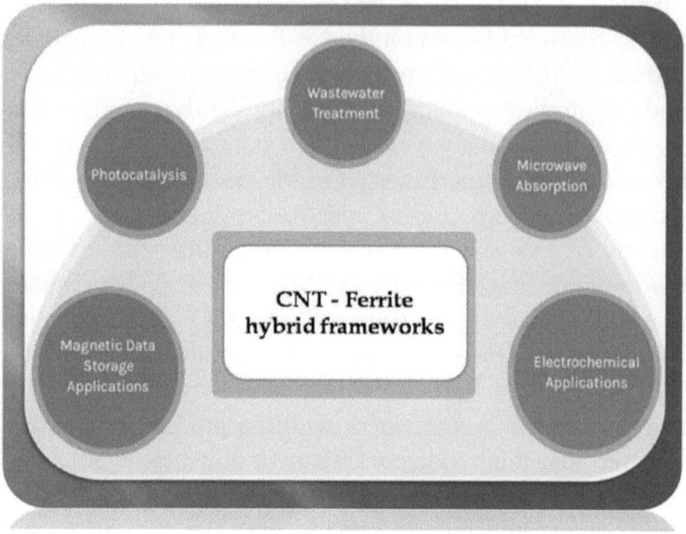

FIGURE 3.1 Applications of CNT–ferrite hybrid frameworks.

Nickel ferrite is an important spinel ferrite nanocrystal owing to its high electrical resistivity, chemical stability, electromagnetic performance, and moderate saturation magnetization (Ms).[6,55,101] They have been efficiently synthesized by many methods such as the sol–gel process,[8,20,43] co-precipitation,[48,81] combustion,[65] and hydrothermal synthesis.[98,112,118] (Komarneni, 1998). $NiFe_2O_4$-based nanocomposites have a wide range of applications like microwave absorbers[85,92] and DNA separation[54] due to their distinctive properties, and CNTs are considered to be the most appropriate for nanocomposite synthesis owing to their nanoscale dimensions and outstanding properties.

The hydrothermal method of synthesis of carbon nanotubes/nickel ferrite ($CNT/NiFe_2O_4$) hybrid nanocomposites was reported by Jin et al.[34] Analytic grade reagents and multiwalled CNTs (MWCNTs) were used in the synthesis of $CNT/NiFe_2O_4$ nanocomposites. Nitric acid was added to pristine CNTs and dispersed ultrasonically, then 0.1 mmol of $Fe(NO_3)_3.9H_2O$ and 0.05 mmol of $Ni(NO_3)_2.6H_2O$ were added to this

CNT suspension. But MWCNTs have poor solubility and chemical inertness due to their structure. To overcome this limitation, the CNTs were functionalized using concentrated aqueous nitric acid and after correcting the pH by adding NaOH solution dropwise, it was transferred to a PTFE-lined stainless-steel autoclave and $NiFe_2O_4$ nanoparticles were hydrothermally synthesized in situ on the CNT surface on an alkaline medium which lasted for about 12 hours. After cooling, the obtained products were separated by filtration and then washed thoroughly to remove any impurities. The morphological analysis of the obtained products was done using transmission electron microscopy (TEM), x-ray diffractometry (XRD), energy-dispersive x-ray (EDX) spectroscopy, and x-ray photoelectron spectroscopy (XPS), and the magnetic properties were examined using a vibrating sample magnetometer (VSM) over an applied magnetic field.

The TEM images clearly displayed the $NiFe_2O_4$ being supported on the CNTs and the XPS results pointed out that the oxygen-containing functional groups were introduced on the CNT surface through nitric acid treatment which was in agreement with the preceding results.[102] It was also observed that this oxidative treatment aided the whole synthesis as they improved the hydrophilicity of the CNTs allowing them to be well dispersed in water and this, in turn, favored the uniform distribution of the synthesized nanoparticles on the CNTs. It was also observed that these functional groups were able to absorb metal ions in solution by electrostatic interactions. Furthermore, the VSM measurements indicated that the CNT/$NiFe_2O_4$ nanocomposites are typical soft ferromagnetic materials having a saturation magnetization value (M_s) of around 24 emu/g. Thus, the CNT/$NiFe_2O_4$ nanocomposites were synthesized using a simple hydrothermal process and confirmed using TEM, XPS, XRD, and EDX which confirmed that $NiFe_2O_4$ was synthesized in situ on the CNT surface.

3.3 APPLICATIONS OF CNT–FERRITE NANOCOMPOSITES

3.3.1 MICROWAVE ABSORPTION

3.3.1.1 MICROWAVE ABSORPTION BY CNT/STRONTIUM FERRITE NANOCOMPOSITE

The microwave absorption properties of strontium ferrite/CNT nanocomposites were significantly enhanced by carbon nanotubes (CNT) and reported by Ali Ghasemi.[21] CNTs are characterized by fascinating properties.[73,58,103] Magnetic multiwalled carbon nanotube (MWCNT) nanocomposite was prepared by creating a strontium ferrite film on the surface of MWCNTs and the structural and magnetic properties of the nanocomposite were evaluated. The synthesis of nanocomposite involves two steps. The first step involved the synthesis of Strontium ferrite through the sol–gel process. In the second step CNTs functionalized in nitric acid and dispersed in ethylene chloride at a constant pH of 10 were added to ferrite solution with ultrasonication. For evaluating the effect of CNT fraction on the properties of the nanocomposite, samples with a different volume percentage of CNT were prepared.

Transmission electron microscopy (TEM) was used to study the morphology of nanoparticles. The TEM image of the nanocomposite synthesized revealed that strontium ferrite thin film was formed on CNTs. To test the structural stability of the CNT–ferrite, it was redispersed in water, sonicated, and then recovered. The thickness of the thin film on CNTs was also found to be uniform. The x-ray diffraction pattern of nanocomposite had the peaks of CNT and Strontium ferrite which reflected that nanocomposite was well formed by the incorporation of CNTs in Strontium ferrite containing solution.

Magnetic properties of nanocomposite were analyzed using Vibrating Sample Magnetometer (VSM). It was revealed from VSM graphs that with an increase in carbon nanotube content the saturation magnetization along with coercivity decreases. The plot of variation of reflection loss with frequency indicates that addition of carbon nanotube to the nanocomposite could enhance the reflection loss values in the whole frequency range. Magnetic studies indicated that the nanocomposites exhibit excellent potential applications in manipulation and organization of magnetic structures for microwave absorption as well as in the field of electromagnetic wave absorbtion.

3.3.1.2 ELECTROMAGNETIC WAVE ABSORBING PROPERTIES OF CNT/GRAPHENE/$BaFe_{12}O_{19}$ COMPOSITE

Materials with electromagnetic wave-absorbing properties are gaining large importance in many fields such as military arms, mobile phones, and electronic gadgets.[23] Zhao et al. synthesized a nanocomposite from

Amorphous Carbon Nano Tube (ACNT), Reduced Graphene Oxide (RGO), and $BaFe_{12}O_{19}$(BF) and evaluated its ability to absorb electromagnetic radiation and related properties.[115] Hummers' method was employed to synthesize Graphene Oxide (GO). 1g Natural flake Graphite and Na_2SO_3 are mixed in conc. H_2SO_4 and stirred in an ice water bath using a magnetic stirring apparatus during low-temperature stage. 6g of Potassium Permanganate was added to reduce the heat influence resulting from the redox reaction. The solution turns dark green when maintained in the same temperature for one hour. The beaker is then transferred to 35°C for middle-temperature stage and is stirred for an hour. Then for high-temperature stage, 80 ml of distilled water was slowly added to the solution and the temperature was slowly raised to 95°C. After 30 minutes, 200 ml of distilled water and 6 ml H_2O_2 were added. The resulting luminous yellow sample is dried after being washed. Barium ferrite (BF) is synthesized from $Fe(NO_3)_2$ and $Ba(NO_3)_3$ mixed with citric acid through self-propagating combustion. The interlinked ACNT/RGO/BF nanocomposite is prepared using the self-propagating combustion method. Oxidized ACNTs, GO, and citric acid mixed with $Fe(NO_3)_3$ and $Ba(NO_3)_3$ are used by dissolving in water and pH is made 7 by the addition of ammonium hydroxide. The colloidal sol produced, when continuously heated and stirred at 350°C, produces ACNT/RGO/BF composite. GO gets reduced by the intense heat during the process.

Scanning electron microscopy (SEM) is used to gain the understanding of the structure of nanocomposite in which ACNTs and RGOs were dispersed in a homogenous manner inside BF forming a conducting interlinked structure, leading to an increase in dielectric loss of composite. Energy-dispersive X-ray spectroscopy (EDS) analysis confirms the presence of all elements and indicates the least amount of Ba and largest amount of Fe. Elemental mapping indicated the uniform mixing of ACNT, RGO, $BaFe_{12}O_{19}$ during the reaction. X-ray diffraction pattern and Raman spectroscopy were used for the microstructure characterization. Transition electron microscopy (TEM) images of ACNT, acidified ACNT, graphene, nanocomposite etc. were obtained. The schematic plot of nanocomposite portrayed the intertwined network which raised the structural stability and electrical conductivity which contributed to electromagnetic wave-absorbing properties.

On analyzing the permeability, permittivity etc. the nanocomposite displayed good dielectric loss and magnetic loss. The reflection loss peak of nanocomposite was found to be −19.03 dB at a frequency of 11.04 GHz in the frequency range of 2 to 18 GHz. For reflection loss below −10.03 dB the frequency bandwidth was 3.8 GHz. The good electromagnetic wave-absorbing properties of the nanocomposite can be attributed to the complex network structure, the skin effect, the multiple reflections and scatterings happening inside dense structures,[78,99] high conductivity of structures leading to energy dissipation as heat etc.

3.3.1.3 MICROWAVE ABSORPTION PROPERTIES OF MWCNTs/NICKEL ZINC FERRITE NANOCOMPOSITES

The introduction of CNT into ferrites improves the microwave-absorbing properties and rise in conductivity of the nanocomposite.[21,32] Development of multiwalled carbon nanotubes (MWCNTs) and $Ni0.5Zn_{0.5}Fe_2O_4$ nanocomposite has been demonstrated by Mustaffa et al.[53] Nickel Zinc Ferrite ($Ni_{0.5}Zn_{0.5}Fe_2O_4$) is employed in electromagnetic wave absorption in a high-frequency region due to its properties. It is used in a radar absorber for aircrafts etc. due to its high density etc. (Gang et al., 2011). It exhibits better microwave absorption properties, compared with other ferrites.[10,96] MWCNT was synthesized through Chemical Vapor Deposition (CVD) method with ethanol as a carbon source and sintered Nickel Zinc Ferrite as a catalyst. Nanocomposite of MWCNT with Nickel Zinc Ferrite was prepared by blending MWCNT and Nickel Zinc Ferrite powder with an epoxy resin in a 60:40 weight ratio. Samples of 3-mm and 2-mm thickness were synthesized. Microstructural, magnetic, and microwave adsorption properties of the nanocomposite were investigated.

Phase and microstructural evaluation of the synthesized sample was performed using X-ray diffraction. The X-ray diffraction pattern confirmed the formation of a single-phase cubic of Nickel Zinc Ferrite and that the structure of MWCNT was retained without any changes. The field emission scanning electron microscopy (FESEM) image of the composite displayed MWCNTs consistently deposited along the outer surface of Nickel Zinc Ferrite. Energy-dispersive X-ray (EDS) spectra revealed the chemical composition of the composite by confirming the presence of Zn, Ni, C, Fe, and O without any other contaminating constituent.

Electromagnetic analysis of the samples was done by permeability and permittivity analysis. Dielectric loss tangent and magnetic loss tangent were demonstrated and it was understood that the composite had a higher value of dielectric loss tangent in the frequency range of 8–12 Hz. This shows that the electromagnetic wave absorption is mainly by the dielectric loss mechanism, not by the magnetic loss mechanism. On evaluating the microwave absorption using reflection loss, reflection loss peak was observed with a reflection loss of −19.34 dB at a frequency of 8.46 GHz for the sample with 3-mm thickness. For a reflection loss below −10 dB the bandwidth obtained was 1.24 GHz. Hence, the MWCNT/$Ni_{0.5}Zn_{0.5}Fe_2O_4$ nanocomposite was found to be a suitable candidate for microwave absorption. The enhancement of microwave absorption in Nickel Zinc Ferrite by MWCNT has been clearly illustrated in this work.

3.3.1.4 CNT/$NI_{O.5}ZN_{0.5}Fe_2O_4$ COMPOSITES AS MICROWAVE ABSORBERS

The development of carbon nanotubes (CNT)/nickel zinc ferrite ($Ni_{0.5}Zn_{0.5}Fe_2O_4$) nanocomposites and their electromagnetic properties has been reported by Zhou et al.[119] Nickel zinc ferrite (NZF) exhibits high resistivity, high Curie temperature, low dielectric losses, and excellent microwave-absorbing properties, and is used in both low- and high-frequency devices that play a very important role in applications such as microwave equipment and rod antennas. One of their most demanding applications in particle accelerators is to serve as a microwave absorber or an on-beam-line higher-order mode (HOM) load. For functioning as a microwave absorber in such a device, the ferrite must be physically and chemically stable at cryogenic temperatures, radiation-tolerant, have wide range bandwidth microwave absorption, good thermal conductivity, and an appropriate DC electrical conductivity for charge–discharge. But their DC electrical conductivity rapidly decreases at low temperatures which was then solved by incorporating a conductive filler to form a conduction network, and one-dimensional CNTs were found to be the most suitable for the same. CNTs have superior electrical property when compared with copper wires and it was found that when they were incorporated with NZF, the electrical conductivity of the composite increased by about 5 orders of magnitude at room temperature. The CNT/NZF nanocomposites were synthesized by in situ chemical precipitation and hydrothermal processing and then sintered using microwave sintering technology. This technique could reduce grain growth and improve the densification rate of ceramics which in turn could lower the sintering temperature and annealing time.

In situ chemical precipitation, hydrothermal processing, and microwave sintering technology was adopted for the development of carbon nanotube (CNT)–$Ni_{0.5}Zn_{0.5}Fe_2O_4$ powders. The multiwalled CNTs (MWCNTs) were oxidized and cooled to room temperature. The oxidized mixture was then separated by filtration, followed by mixing with sodium lignosulfonate (SLS) and ultrasonication treatment from this mixture, the excessive SLS was removed, and the obtained solution was well dispersed and stored in deionized water. The CNT suspension so obtained was mixed with a solution containing nickel, zinc, and ferrite in the molar ratio 0.5:0.5:2, and this mixture solution was added dropwise to NaOH solution till the pH reached 10.5. The precursor was then placed in a Teflon-lined autoclave and treated hydrothermally. The obtained precipitates were finally washed, dried, cold-pressed, and then microwave sintering (MWS) technology adopted to consolidate the samples. The Archimedes technique was used to measure the density of the sintered samples, and the physical properties measurement system was used for the measurement of electrical conductivity and saturated magnetization. The scanning electron microscopy (SEM) was used for the morphological analysis of the prepared composites and transmission electron microscopy (TEM) for the study of microstructures. The X-Ray Diffractometry (XRD) of the composite powders was also taken which indicated that the incorporation of CNTs had no considerable impact on the formation of the designed ferrite phase. The study concluded that CNT-doped ferrite ceramics are suitable to function as HOM load due to their excellent low-temperature characteristics.[119]

3.3.1.5 MAGNETIC AND REFLECTION LOSS CHARACTERISTICS OF CNT/SUBSTITUTED STRONTIUM FERRITE NANOCOMPOSITES

The magnetic and reflection loss characteristics of Strontium ferrite and their nanocomposites with Carbon Nano Tubes (CNTs) were evaluated by Ghasemi et al. Nanocomposites were synthesized by hetero-coagulation. High Resolution Transmission Electron Microscopy (HRTEM) was employed to analyze the structure of the composite and it was confirmed that the ferrite nanoparticles were successfully attached on the CNT surface. Magnetic properties were evaluated using Vibrating Sample Magnetometer and the effect of volume percentage of CNTs on saturation magnetization was revealed. Nanocomposites exhibited large reflection losses and this increased with volume percentage of CNTs in them. The microwave-absorbing bandwidth can be varied by varying the thickness of the sample. The nanocomposite was recognized as a potent candidate for the construction of electromagnetic wave absorbers through this study.[22]

3.3.1.6 RADAR ABSORPTION BY $BaFe_{12}O_{19}/ZnFe_2O_4$/CNT NANOCOMPOSITES

Tyagi et al. have reported the synthesis of a magnetic composite consisting of barium hexaferrite and zinc spinel ferrite nanoparticles ($BaFe_{12}O_{19}/ZnFe_2O_4$) by the auto-combustion method.[93] CNTs were synthesized by the thermal decomposition of acetylene gas and a RADAR-absorbing medium was developed by dispersing CNTs in the magnetic composite prepared. Furthermore, the microwave absorption properties of the composite powder consisting of hard magnetic barium hexaferrite and soft magnetic zinc ferrite were measured, and the effect of the dispersion of CNTs in the ferrite matrix was also examined.

Barium hexaferrite ($BaFe_{12}O_{19}$) is an M-type hard magnetic material having high hysteresis attenuation, high saturation magnetization,[67] and is extensively used as a magnetic filler in the radar-absorbing materials (RAM). Zinc ferrite ($ZnFe_2O_4$) is a soft spinel ferrite magnetic material and is widely used as a RAM.[18,92] CNTs[45] are conductive fillers and an ideal microwave-absorbing material owing to their unique physical, chemical, mechanical, and electrical properties.[117] Solid-state reaction method is a conventional way of synthesizing ferrite nanoparticles in which the resulting powders are of grain size/coarse particles.[12] So, synthesizing a uniform powder without impurity is not possible with this method. In the auto-combustion synthesis method, the solution combustion occurs at a lower flame temperature with higher composition homogeneity, involving both exothermic and thermally induced anionic redox reaction of xerogel. In this method, an enormous amount of gas is evolved which results in highly voluminous and foamy powders with a large specific area and soft agglomeration.[29] This method is economical, has a high production rate, and is an ideal substitute for the solid-state reaction method. CNTs were synthesized using the chemical vapor deposition (CVD) method from which high-quality multiwalled CNTs (MWCNTs) were synthesized at relatively low cost.[88]

All the chemicals used for the synthesis of the nanocomposite were of AR/GR grade. For the synthesis of the $BaFe_{12}O_{19}/ZnFe_2O_4$ magnetic composite, a stoichiometric amount of metal nitrates was dissolved into pure water (solution I) and barium carbonate into acetic acid (solution II). Then solutions I and II were mixed with citric acid and the pH of the resulting solution was raised to 7.5 by the addition of ammonia solution. The resulting sol was then heated at a constant temperature, evaporated and after the removal of the solvent, the dried precursor underwent a self-ignition reaction to form the nanocrystalline ferrite. The MWCNTs were grown over silicon substrate using the CVD method and were dispersed in the ferrite matrix to form the $BaFe_{12}O_{19}/ZnFe_2O_4$/CNTs nanocomposite. Then x-ray diffractometer (XRD) was used to identify the crystalline phase of the annealed ferrite samples and CNTs, and the field emission scanning electron microscope (FESEM) for examining the morphological features of the prepared nanocomposite. The magnetization study of the ferrite was done using vibrating sample magnetometer (VSM) and the reflection loss measurements were taken out on Network Analyzer using material measurement software. The Raman spectra of the CNTs were also recorded at room temperature using the Renishaw Raman imaging microscope.

It was found that the crystalline size of the $ZnFe_2O_4$ phase and the reflection loss of the magnetic composite increased with an increase in heat treatment temperature.[92] Furthermore, the XRD patterns of CNTs showed that their interlayer spacing is slightly higher when compared with pure graphite due to the influence of the Fe catalyst used in their synthesis. Furthermore, the VSM pattern of the magnetic composite showed low coercivity and low remanence values in the "as-synthesized" condition which confirmed the formation of the soft magnetic phase, which in turn agrees with the XRD pattern. Moreover, the reflection loss and the absorption bandwidth of the developed magnetic composite were found to be higher than pure ferrite nanoparticles. The microwave absorption performance of the prepared nanocomposite was also measured and it confirmed that the $BaFe_{12}O_{19}/ZnFe_2O_4$/CNTs nanocomposite exhibited superior microwave absorption when compared with pure $BaFe_{12}O_{19}/ZnFe_2O_4$ and pure CNTs owing to the diamagnetic nature of CNTs and ferromagnetic nature of $BaFe_{12}O_{19}/ZnFe_2O_4$ nanoparticles.[97] So, it was concluded that the reinforcement of a certain mass of CNTs on the $BaFe_{12}O_{19}/ZnFe_2O_4$ magnetic composite improved their RADAR absorption properties.

3.3.2 WASTE WATER TREATMENT

3.3.2.1 ADSORPTION OF RHODAMINE B FROM AQUEOUS SOLUTIONS

Oyetade et al. analyzed the effectiveness of nanocomposites prepared from Carbon Nano Tubes (CNTs) and cobalt ferrite in the process of adsorption of the commonly used dye, rhodamine B (RhB) from aqueous solutions.[60] RhB that gets discharged into water bodies is causing great environmental hazards and health hazards for living community. Hence, it is necessary to think of potent methods for the removal of this dye from water. Among various methods such as coagulation,[110] irradiation,[104] photochemical processes,[62] decoloration using white fungi[52], adsorption was found to be the most feasible and eco-friendly one.[25] CNTs were found to be super-efficient in adsorbing pollutants[61] and magnetic nanocomposites of CNTs helped in overcoming various other limitations. The efficiency of a nanocomposite of CNT with cobalt ferrite is analyzed here.

Cobalt ferrite nanoparticles were prepared from stable salts of cobalt and iron by the wet chemical or coprecipitation method. The advantage of using this method over the sol–gel process and various other methods such as evaporation condensation commonly used in producing nanoparticles is that it is relatively easy to control the production, the particle size etc. and extra microwave, as well as mechanical treatments, are not required.[49] Pristine multiwalled carbon nanotubes (MWCNTs) were made free from impurities and then functionalized in a mixture of concentrated sulfuric acid and nitric acid. Functionalized MWCNTs were treated with solutions of stable salts of iron and cobalt and nanocomposites of varying percentages were prepared.

Batch adsorption experiments were performed to compare the efficiency of adsorbents. 50 mg of adsorbent was added to 25 cm^3 of freshly prepared RhB solutions of known concentrations and pH values. The mixtures were maintained at a fixed temperature for 24 hrs and were filtered by gravity. Ultraviolet-visible spectrophotometry was employed to determine the RhB concentration in the filtrate. The optimum conditions for adsorption were evaluated by studying the effect of factors such as pH, initial concentration of dye, and temperature.

The presence of functional groups in MWCNTs was confirmed using Fourier transform infrared (FTIR) spectrometry. The FTIR spectrum also helped in confirming the successful synthesis of the nanocomposite. Transition Electron Microscopy (TEM) and High-Resolution Transition Electron Microscopy (HRTEM) images were used to observe the structure and size distribution of the samples. The successful synthesis of nanocomposites that preserved the tubular structure of CNTs and ferrites with cubic structure was once again confirmed.

Raman spectroscopy was employed to evaluate the defects and crystallinity of MWCNTs, and it was found that the disorder was more in functionalized MWCNTs. This could be due to the oxidation of MWCNTs. The nanocomposites were also found to have a larger surface area and an increase in functionalized MWCNT content in the nanocomposite resulted in a further increase in area, pore volume, and pore diameter.

The effect of pH of adsorbate on adsorption was analyzed by observing adsorption in solutions of pH from 1 to 10. Adsorption was found to be low in low values of pH and then reached a maximum between pH of 6–8 for all adsorbents (nanocomposites of different percentage, MWCNT, cobalt ferrite). Increase in basicity reduced adsorption as the dyes were modified at higher pH values.[114] MWCNT was recognized as the best adsorbent and the increase in adsorption with the increase in MWCNT content in nanocomposite was noticed. On analyzing the effect of contact time, equilibrium was found to be established in 6 hours. The kinetics of adsorption was found to be the best suiting to that of a second-order model. Increase in adsorbent also enhanced adsorption as an increase in mass leads to a larger surface area for adsorption. The temperature rise also improved adsorption indicating the endothermic nature of the process. But for 50% nanocomposites, it was found to be exactly opposite. Langmuir isotherm explained the equilibrium conditions of adsorption in the best way indicating monolayer adsorption on the homogenous surface of the adsorbent. On evaluating the thermodynamic parameters, the process was found to be spontaneous and feasible with negative Gibbs energy. It was found to be entropy driven expect for 50% nanocomposite.

Desorption studies performed using Ethanol and Acetone proved the reusability of the nanocomposite. Acetone was found to be more efficient in desorption. The study by Oyetade et al. thus demonstrated the potential of the CNT–cobalt ferrite nanocomposite in dye removal.

3.3.2.2 ADSORPTION OF ARSENIC (V) IONS FROM WASTE WATER

Ahangari et al.[2] employed spinel ferrites for the removal of arsenic ions which is a nonbiodegradable heavy metal that affects the human health drastically by its intake from water contaminated with arsenic[83] and their poisonous variants present in industrial wastewater are mostly in the form of inorganic arsenic species. For the removal of arsenic from industrial wastewater, the Environment Protection Agency (EPA) has approved many methods such as reverse osmosis,[3] electrochemical treatment,[59] chemical precipitation,[35] and ion exchange.[86] But these methods did not prove to be completely effective in the removal of arsenic effluents from wastewater, and it was found that the adsorption method whose efficiency strongly depends on the adsorbent characteristics was the most suitable one for the same. Moreover, the adsorbents can be regenerated by the process of desorption.[69] Spinel ferrites such as the nickel zinc ferrite (NZF) act as able adsorbents.[66] Nevertheless, their direct usage has certain drawbacks, and so it is essential to lay the Ferrite NPs on a substrate such as natural minerals, concrete,[28] activated carbons,[1] and CNT to enhance its adsorption. In this work, the inverse coprecipitation method was employed for the preparation of NZF ($Ni_{0.5}Zn_{0.5}Fe_2O_4$) and its nanocomposite with carbon nanotubes CNZF ($Ni_{0.5}Zn_{0.5}Fe_2O_4$/CNT). Furthermore, the parameters such as pH, contact time, and adsorbent dosage were optimized for both the adsorbents, and the kinetic and isotherm models of NZF and CNZF were demonstrated.

For the preparation of NZF, 2 moles of $FeCl_3$ with 0.5 moles of $ZnCl_2$ and 0.5 moles of $NiCl_2.6H_2O$ were added to deionized water and homogenized; 200 ml of aqueous sodium hydroxide solution was added and stirred till it is completely precipitated. The precipitates were then washed thoroughly with distilled water and ethanol and dried in an oven to produce NZF.[24] For the preparation of CNZF, 10 wt.% of MWCNTs with the functional group of COOH were dispersed in 100 ml of ethanol; the metal solution was added to this dispersed solution and the NZF nanoparticles were synthesized homogenously on the surface of MWCNTs by the inverse coprecipitation method.[26,48] Then the phase constitution of the obtained samples was examined using the X-ray diffractometery (XRD) and the mean size of the NZF and CNZF samples was obtained using the Scherrer equation. In the course of the experiment to study the adsorption of the arsenic ions, the samples were stirred with simulated arsenic solutions and after the adsorption, the samples were separated from the supernatant using a magnet. The solutions were then studied using inductively coupled plasma atomic emission spectroscopy and the concentration of the arsenic ions left in the solution was estimated. The amount of adsorbed arsenic ions per solution was obtained using the equation[31]

$$q_e = \frac{(C_0 - C_e)v}{m}$$

where $\boldsymbol{q_e}$ is the adsorption capacity of the adsorbent, $\boldsymbol{C_0}$ is the concentration of the anions under the initial condition, $\boldsymbol{C_e}$ is the concentration of the anions at the equilibrium condition, $\boldsymbol{m}$ is the weight of sorbent (g), and $\boldsymbol{V}$ is the volume of the solution.

The XRD (X-Ray Diffraction), FTIR (Fourier Transform Infrared Spectroscopy), FESEM (Field Emission Scanning Electron Microscope), and XPS (X-ray photoelectron spectroscopy) were used to confirm the production of novel magnetic adsorbents. It was seen that the pH of the solution had a major influence in the arsenic ion adsorption process and since the arsenic ions had a negative charge, the maximum positively charged point of the adsorbents which came about at a pH of 2 was found to be the optimal condition for maximum adsorption.[50] Also, the adsorption of arsenic by both the adsorbents was found to increase considerably with the increment of CNZF and NZF adsorbent dosage and a high adsorption rate was observed for a contact time up to 5 minutes for both the adsorbents. Since the CNZF had a greater specific surface area when compared with NZF, they showed maximum adsorption capacities. The Langmuir adsorption isotherm successfully simulated the adsorption isotherms of arsenic ions by NZF and CNZF and the pseudo-second model was selected as the optimal kinetic model for studying the kinetic behavior of adsorption over time. This experiment concluded that the NZF and CNZF powders were effective in removing arsenic (V) ions from contaminated wastewater.

3.3.3 ELECTROCHEMICAL APPLICATIONS

3.3.3.1 BIFUNCTIONAL NICKEL FERRITE DECORATED CNTs AS AIR CATHODE OF Zn–AIR BATTERIES

Metal–air battery being the modern promising energy technologies, great efforts have been made to enhance the efficiency of batteries especially of Zn–air. Cost-effective bifunctional air electrode has a notable role in advancing the practical application of rechargeable metal–air batteries. Integrating advanced bifunctional oxygen electrocatalysts to make highly efficient air electrode is of great benefit for the improvement of battery performance.[61] It gives ultra-stable cyclability over 1500 h at 5 mA cm^{-2} and a large peak power density of 194 mW cm^{-2}. Vertically Aligned Carbon Nanotubes (VACNTs) that combine the function of current collector, supporter, and catalyst have high potential for the construction of a three-dimensional free-standing air electrode.[7,33,56] They construct an integrated novel free-standing air electrode by a developed post annealing method assisted by supercritical carbon dioxide ($SCCO_2$). This hybrid consists of oxygen vacancy-rich $NiFeO_x$ that acts as an oxygen electro catalyst of high efficiency and VACNTs that provide essential pores/channels and active framework ($NiFeO_x$@VACNTs). This exhibits the outstanding stability and catalytic activity. Scanning electron microscopy (SEM) image of $NiFeO_x$@VACNTs gives rougher surface in comparison with earlier VACNTs. Above this, carbon nanotubes of vertical alignment are crosslinked with each other. Transmission electron microscopy (TEM) images add that the growth of these particles occurs on a rougher surface with a size of 10 nm and the nanoparticles show a 0.29-nm lattice fringe width. Investigation of structure width is made by Raman spectra. Loading amount of $NiFeO_x$ is found to be 20 wt% by thermo gravity analysis. Nitrogen adsorption/desorption results point out that this hybrid has a larger Brunauer-Emmett-Teller (BET) surface area and a great porous structure that has a role in the improvement of electrochemical performance of Zn–air batteries. Increasing the annealing temperature results in the weakening of the diffraction peaks. Considering the excellent bifunctional OER/ORR activity of $NiFeO_x$@VACNTs, a comparison is made. Zn–air battery driven by $NiFeO_x$@VACNTs has higher stable open circuit voltage (OCV), superior current density, peak power density, excellent rechargeable capability, and higher specific capacity in comparison with

Pt/C+TrO_2, $NiFeO_x$/VACNTs mixture, and pristine VACNTs. This work provides a way for the development of a cost-effective and free-standing integrated air cathode of Zn–air batteries and related technologies.

3.3.4 MAGNETIC DATA STORAGE APPLICATIONS

3.3.4.1 MAGNETIC PROPERTIES OF $CoFe_2O_4$ NANOWIRES/CNT NANOCOMPOSITE

Cobalt ferrite ($CoFe_2O_4$) is one of the most studied magnetic nanomaterial.[72] Magnetic nanowires with extremely small diameters are extracted by nonmagnetic materials to reduce magnetization polar relaxation and to increase coercivity (Mekala, 2000). Synthesis of $CoFe_2O_4$ nanowires with an average diameter of 50 nm and length of some micrometres were encapsulated inside CNTs using the confinement effect by the surrounding nanotubes.[63] Cap formation near tips of tube shows nanotube itself considered as a closed nanoreactor. That is, reaction condition and macroscopic conditions outside the tube are different. The introduction of foreign elements such as V_2O_5, Ag, Fe, Cr, Nb, and Cd is carried out through methods of wet chemical solution or by which capillary forces induce the filling by molten material.[15] These methods require great temperature (>7000 C). With both filled and externally coated carbon nanotubes being in the final sample, it is to be noted that filling selectively remained relatively low in comparison with external coating. Enhancement of surface relaxation causes the difference in magnetic properties in nanoscale and bulk material. $CoFe_2O_4$ synthesis was done by filling CNTs by an aqueous solution of iron and cobalt. Powder X ray diffraction (XRD) was used for checking the crystallinity. Microstructural characterization was made by transmission electron microscopy (TEM). Mossbauer spectra were recorded after the calcination to quantify the amount of Fe in each crystallographic site and to detect the presence of a disordered phase. Increase in calcination temperature leads to a dramatic microstructural change. The porous structure changes to well-crystallized nanowire. Nanowires of 25 nm diameter with high coercivity were obtained at room temperature by submitting the $CoFe_2O_4$ nanowires after calcination in air at 1000°C to an argon treatment at 5500°C for 2 hours. Irrespective of the drawback of particle agglomeration and magnetic loss by dipolar relaxation, this magnetic nanowire finds a wide range of applications in high-density data storage.

3.3.5 PHOTOCATALYSIS

3.3.5.1 PHOTOCATALYTIC PROPERTIES OF Co-DOPED COPPER FERRITE/CNTs NANOCOMPOSITES

Hybrid nanocomposite of carbon nano tubes (CNTs) with cobalt-doped copper ferrite nanoparticles was developed by Singh et al and its properties were analyzed.[76] The structural, magnetic, optical, and photocatalytic properties of nanocomposites of Carbon Nano Tubes (CNTs) with cobalt-doped copper ferrite nanoparticles were studied and compared with that of individual nanoparticles.

The microemulsion synthesis method was employed to dope Cobalt in Copper Ferrite and this was encrusted on the curved of CNTs. This resulted in complete stabilization of nanoparticles. The low-cost microemulsion method reduced the issues such as agglomeration, irregular size distribution, and inadequate attachment etc. which arises when other techniques like polymer wrapping method, solvothermal method, pot polyol method etc. are used. Here, two microemulsion systems were prepared using water: sodium dodecyl sulfate: hexane: 1-butanol in an appropriate weight ratio, and stirred for 15 min until the solutions became transparent. Metal salts were dissolved in system one, 20 ml precipitating agent (sodium hydroxide) in system two, and two systems were slowly mixed with continuous mixing. Brown colored precipitate obtained was filtered, washed, and dried to yield stable nanoparticles. Multiwalled carbon nano tubes (MWCNTs) were synthesized by the arc discharge process, and washed, dried, and functionalized with nitric acid. These CNTs were dispersed in microemulsion system one with metal salt solution. This was mixed with system two in which 5 M NaOH was dissolved with continuous stirring. The precipitate was washed, purified, and dried to obtain the desired nanocomposite.

Fourier transformation infrared (FTIR) spectra of the composite revealed the M-O bond due to the presence of metal ion in tetrahedral sites and the attachment of nanoparticles with CNTs was confirmed. The formation of cobalt-doped copper ferrite nanoparticles and their nanocomposites were analyzed and confirmed using the powder x-ray diffraction method. Particle size was found to be 4–5 mm using High Resolution Transmission Electron Microscopy (HRTEM). The formation of a fine layer of nanoparticles on the surface of CNTs was also confirmed. From studies employing vibrating sample magnetometer a significant rise in coercivity and saturation magnetization was observed. This is due to doping with cobalt. Wide coverage of visible spectrum was understood from the optical studies made. When evaluated using Rhodamine B dye, the nanocomposite exhibited higher photocatalytic activity.

3.3.5.2 PHOTOCATALYTIC PERFORMANCE OF $Co_XNi_{1-X}Fe_2O_4$/MWCNTs NANOCOMPOSITES

Photocatalytic technology is an environment-friendly, highly efficient, and reusable technology[51,57] which has been recently developed and is new in the field of water treatment. $Co_xNi_{1-x}Fe_2O_4$/multiwalled carbon nanotube nanocomposites have been proved to be efficient photocatalysts. Lu et al. have reported the fabrication of CNF/MWCNTs nanocomposites[46] and investigated the photodegradation efficiency of the prepared photocatalysts in the presence of hydrogen peroxide (H_2O_2). Moreover, the prepared composites were characterized using various analytical methods.

Spinel ferrites are semiconductor compounds that have a narrow band gap, good photosensitivity, and stable photoelectrochemical performance (Bhasker et al., 2018) (Sun et al., 2010).[27,106] When irradiated by light, it produces photogenerated electron–hole pairs that in turn can initiate redox reactions to achieve photocatalysis[17,87] (Liu and Li, 2018) (Aji and Suharyadi, 2017) (Anjana et al., 2018). Furthermore, they can catalyze the decomposition of hydrogen peroxide producing hydroxyl radicals which in turn can oxidize organic pollutants such as synthetic dyes in water.[19,68] Lately, researches are paying more attention to further improving the adsorption capacity and promoting the separation rate of photogenerated electron–hole pairs to broaden the applications of photocatalysts considerably[82,111] (Silambarasu et al., 2017) (Ketolainen et al., 2015). CNTs are one-dimensional nanomaterials that have outstanding properties that make them very useful in a wide range of applications such as hydrogen storage, electronic devices, supercapacitors, and batteries (Nawale et al., 2017).[76,113,116] Carbon materials on exposure to UV light can produce free radicals, improve the catalytic performance, and aid in the degradation of organic pollutants through the interaction of carbon and light.[40,74] Hence, assembling CNF on MWCNTs improves the overall catalytic performance and applications in environmental remediation.[47,100] The reagents used for the same were of analytical grade and the MWCNTs had a purity of more than 95 wt%. The MWCNTs were thoroughly dispersed in a mixture of concentrated sulfuric acid and nitric acid and heated in a pressure reactor, so as to optimize the reaction conditions. The prepared products were then centrifuged, washed, and oven-dried.

Lu et al. have reported the photocatalytic performance of $Co_xNi_{(1-x)}Fe_2O_4$/multiwalled carbon nanotube nanocomposites.[46] First, MWCNTs, glycol, and $CO(NH_2)_2$ were added to the PTFE reactor and hydrothermally treated in an oven for about 5 hours. Once the reaction was complete, the autoclave was cooled and the supernatant was washed thoroughly with desalted water and ethanol. The autoclave was then dried and the sample grounded using a mortar. The morphological analysis of the photocatalysts was done using scanning electron microscopy (SEM), transmission electron microscopy (TEM), and high-resolution transmission electron microscopy (HRTEM) which confirmed that the MWCNTs were completely coated with CNF and revealed its polycrystalline structure. The crystallization and specific area of the samples were examined using X-ray diffraction (XRD) and volumetric gas sorption instrument, respectively. The optical properties were studied using a UV-2550 spectrophotometer and Raman spectroscopy was also done with laser light. Finally, methylene blue (MB) was used as a research object to evaluate the catalytic effect of the catalyst in the presence of H_2O_2. Moreover, the nanocomposite could be easily separated from the dye wastewater by magnetic separation techniques which made the separation process very easy. It was also observed that the degradation of the dyes was lower in acidic and neutral solutions.

The report concluded that all the carbon-based nanocomposites showed superiority over pure ferrites in photocatalytic performances. Using the nanocomposites shortened the reaction time and enhanced the removal rate. It was also observed that there was no significant decrease in photodegradation efficiency after reuse, suggesting that CNF/MWCNTs are recyclable and have enormous potential for the photodegradation of organic pollutants. So, their results indicated that the CNF/MWCNT nanocomposite can be effectively used for the removal of organic pollutants as a photocatalyst.

3.4 CONCLUSION

This chapter gives an overall description of the enhanced performance exhibited by the CNT–ferrite hybrid nanocomposite for various applications owing to their unique and enhanced properties. The various synthesis methods of the ferrite–CNT nanocomposites, such as coprecipitation and other synthesis routes, with a special emphasis on the hydrothermal method of synthesis, are described. The potential of strontium ferrite/CNT nanocomposite, CNT/graphene/$BaFe_{12}O_{19}$ composite, MWCNTs/nickel zinc ferrite nanocomposite, and CNT/substituted strontium ferrite nanocomposites for microwave absorption is reported in detail. The ferrite–CNT hybrid nanocomposites are hence recognized as potential microwave absorbers. The potential of CNT/cobalt ferrite nanocomposite in the removal of Rhodamine B dye from aqueous solution and that of CNT/nickel zinc ferrite in the adsorption of arsenic (V) ions from wastewater are highlighted. The bifunctional nickel ferrite decorated CNTs were realized as an efficient, cost-effective, and free-standing integrated air cathode of Zn–air batteries. $CoFe_2O_4$ nanowires synthesized within CNTs were investigated for their magnetic properties. The photocatalytic performance of copper ferrite nanoparticles on CNTs and $Co_xNi_{1-x}Fe_2O_4$/MWCNT nanocomposite was examined. It was also proved that the $BaFe_{12}O_{19}/ZnFe_2O_4$/CNTs nanocomposite exhibited superior microwave absorption when compared with pure $BaFe_{12}O_{19}/ZnFe_2O_4$ and pure CNTs. To conclude, this chapter outlines the overall research advancements in the development of hybrid nanocomposites of carbon nanotubes and ferrites for potential applications.

KEYWORDS

- **CNT**
- **ferrites**
- **EMI shielding**
- **water treatment**
- **photocatalysis**
- **magnetic data storage**

REFERENCES

1. Ai, L.; Li, M.; Li, L. Adsorption of Methylene Blue from Aqueous Solution with—Activated Carbon/Cobalt Ferrite/Alginate Composite Beads: Kinetics, Isotherms, and Thermodynamics. *J. Chem. Eng. Data* **2011,** *56* (8), 3475–3483.
2. Ahangari, A.; Raygan, S.; Ataie, A. Capabilities of Nickel Zinc Ferrite and Its Nanocomposite with CNT for Adsorption of Arsenic (V) Ions from Wastewater. *J. Environ. Chem. Eng.* **2019,** *7* (6), 103493.
3. Alnaizy, R.; Aidan, A.; Qasim, M. Copper Sulfate as Draw Solute in Forward Osmosis Desalination. *J. Environ. Chem. Eng.* **2013,** *1* (3), 424–430.
4. Asiri, S.; Sertkol, M.; Güngüneş, H.; Amir, M.; Manikandan, A.; Ercan, İ.; Baykal, A. The Temperature Effect on Magnetic Properties of NiFe2O4 Nanoparticles. *J. Inorg. Organomet. Polym. Mater.* **2018,** *28* (4), 1587–1597.
5. Bogue, R. Nanocomposites: A Review of Technology and Applications. *Assem. Autom.* **2011,** *31* (2), 106–112.
6. Cao, X.; Meng, J.; Mi, F.; Zhang, Z.; Sun, J. Preparation and Magnetic Property Investigation of a Nickel Spinel Ferrite-Coated Tetrapod-Like ZnO Composite. *Solid State Commun.* **2011,** *151* (9), 678–682.
7. Cai, X.; Xia, B. Y.; Franklin, J.; Li, B.; Wang, X.; Wang, Z.; Chen, L.; Lin, J.; Lai, L.; Shen, Z. Free-Standing Vertically-Aligned Nitrogen-Doped Carbon Nanotube Arrays/Graphene as Air-Breathing Electrodes for Rechargeable Zinc–Air Batteries. *J. Mater. Chem. A* **2017,** *5* (6), 2488–2495.
8. Chen, X. G.; Ye, Y.; Cheng, J. P. Recent Progress in Electromagnetic Wave Absorbers. *J. Inorg. Mater*. **2011,** *26* (5), 449–456.
9. Chernozatonskii, L. A.; Kosakovskaja, Z. J.; Fedorov, E. A.; Panov, V. I. New Carbon Tubelite-Ordered Film Structure of Multilayer Nanotubes. *Phys. Lett. A* **1995,** *197* (1), 40–46.
10. Cho, S. B.; Kang, D. H.; Oh, J. H. Relationship between Magnetic Properties and Microwave-Absorbing Characteristics of Niznco Ferrite Composites. *J. Mater. Sci.* **1996,** *31* (17), 4719–4722.
11. Cho, W. S.; Hamada, E.; Kondo, Y.; Takayanagi, K. Synthesis of Carbon Nanotubes from Bulk Polymer. *Appl. Phys. Lett.* **1996,** *69* (2), 278–279.
12. Cullity, B. D.; Graham, C. D. *Introduction to Magnetic Materials*; John Wiley & Sons, 2011.
13. Ding, J.; Chen, Y. J.; Shi, Y.; Wang, S. High Coercivity in SiO 2-doped $CoFe_2O_4$ Powders and Thin Films. *Appl. Phys. Lett.* **2000,** *77* (22), 3621–3623.
14. Dionne, G. F. A Review of Ferrites for Microwave Applications. *Proc. IEEE* **1975,** *63* (5), 777–789.
15. Dujardin, E.; Ebbesen, T. W.; Hiura, H.; Tanigaki, K. Capillarity and Wetting of Carbon Nanotubes. *Science* **1994,** *265* (5180), 1850–1852.
16. Durmus, Z.; Durmus, A.; Kavas, H. Synthesis and Characterization of Structural and Magnetic Properties of Graphene/Hard Ferrite Nanocomposites as Microwave-Absorbing Material. *J. Mater. Sci.* **2015,** *50* (3), 1201–1213.

17. Fellows, B. D.; Sandler, S.; Livingston, J.; Fuller, K.; Nwandu, L.; Timmins, S.; Lantz, K. A.; Stefik, M.; Mefford, O. T. Extended Lamer Synthesis of Cobalt-Doped Ferrite. *IEEE Magn. Lett.* **2017,** *9*, 1–5.
18. Feng, J.; Hou, Y.; Wang, Y.; Li, L. Synthesis of Hierarchical ZnFe2O4@ SiO2@ RGO Core–Shell Microspheres for Enhanced Electromagnetic Wave Absorption. *ACS Appl. Mater. Interfaces* **2017,** *9* (16), 14103–14111.
19. Fu, Y.; Chen, H.; Sun, X.; Wang, X. Combination of Cobalt Ferrite and Graphene: High-Performance and Recyclable Visible-Light Photocatalysis. *Appl. Catal. B: Environ.* **2012,** *111*, 280–287.
20. George, M.; John, A. M.; Nair, S. S.; Joy, P. A.; Anantharaman, M. R. Finite Size Effects on the Structural and Magnetic Properties of Sol–Gel Synthesized NiFe2O4 Powders. *J. Magn. Magn. Mater.* **2006,** *302* (1), 190–195.
21. Ghasemi, A. Remarkable Influence of Carbon Nanotubes on Microwave Absorption Characteristics of Strontium Ferrite/CNT Nanocomposites. *J. Magn. Magn. Mater.* **2011,** *323* (23), 3133–3137.
22. Ghasemi, A.; Shirsath, S. E.; Liu, X.; Morisako, A. A Comparison between Magnetic and Reflection Loss Characteristics of Substituted Strontium Ferrite and Nanocomposites of Ferrite/Carbon Nanotubes. *J. Appl. Phys.* **2012,** *111* (7), 07B543.
23. Ghzaiel, T. B.; Dhaoui, W.; Schoenstein, F.; Talbot, P.; Mazaleyrat, F. Substitution Effect of Me = Al, Bi, Cr and Mn to The Microwave Properties of Polyaniline/BaMeFe11O19 for Absorbing Electromagnetic Waves. *J. Alloys Comp.* **2017,** *692*, 774–786.
24. Gul, I. H.; Ahmed, W.; Maqsood, A. Electrical and Magnetic Characterization of Nanocrystalline Ni–Zn Ferrite Synthesis by Co-Precipitation Route. *J. Magn. Magn. Mater.* **2008,** *320* (3–4), 270–275.
25. Hamza, I.; Martincigh, B.; Ngila, J.; Nyamori, V. Adsorption Studies of Aqueous Pb(II) Onto a Sugarcane Bagasse/Multi-Walled Carbon Nanotube Composite. *Phys. Chem. Earth A/B/C* **2013,** *66*, 157–166.
26. Hao, Z.; Liu, Q. F.; Wang, J. B. Coating Carbon Nanotubes with Ferrites Using an Improved Co-Precipitation Method. *J. Comp. Mater.* **2010,** *44* (3), 389–395.
27. He, H. Y. Photocatalytic Degradations of Dyes on Magnetically Separable Ni1−x Co x Fe2O4 Nanoparticles Synthesized by a Hydrothermal Process. *Part. Sci. Technol.* **2016,** *34* (2), 143–151.
28. Hernández-Flores, H.; Pariona, N.; Herrera-Trejo, M.; Hdz-García, H. M.; Mtz-Enriquez, A. I. Concrete/Maghemite Nanocomposites as Novel Adsorbents for Arsenic Removal. *J. Mol. Struct.* **2018,** 1171, 9–16.
29. Huang, J.; Zhuang, H.; Li, W. Optimization of the Microstructure of Low-Temperature Combustion-Synthesized Barium Ferrite Powder. *J. Magn. Magn. Mater.* **2003**, *256* (1–3), 390–395.
30. Iijima, S. Nature (London). *Phys. Sci.* **1991,** *354*, 56.
31. Inchaurrondo, N.; Di Luca, C.; Mori, F.; Pintar, A.; Žerjav, G.; Valiente, M.; Palet, C. Synthesis and Adsorption Behavior of Mesoporous Alumina and Fe-Doped Alumina for the Removal of Dominant Arsenic Species in Contaminated Waters. *J. Environ. Chem. Eng.* **2019,** *7* (1), 102901.

32. Jiang, L.; Gao, L. Carbon Nanotubes—Magnetite Nanocomposites from Solvothermal Processes: Formation, Characterization, and Enhanced Electrical Properties. *Chem. Mater.* **2003,** *15* (14), 2848–2853.
33. Ji, D.; Fan, L.; Li, L.; Peng, S.; Yu, D.; Song, J.; Ramakrishna, S.; Guo, S. Atomically Transition Metals on Self-Supported Porous Carbon Flake Arrays as Binder-Free Air Cathode for Wearable Zinc—Air Batteries. *Adv. Mater.* **2019,** *31* (16), 1808267.
34. Jin, X.; Ni, Q. Q.; Fu, Y.; Zhang, L.; Natsuki, T. Hydrothermal Synthesis of Carbon Nanotube/Nickel Ferrite Nanocomposites. *Sen'i Gakkaishi* **2012,** *68* (5), 112–117.
35. Kameda, T.; Suzuki, Y.; Yoshioka, T. Removal of Arsenic from an Aqueous Solution by Coprecipitation with Manganese Oxide. *J. Environ. Chem. Eng.* **2014,** *2* (4), 2045–2049.
36. Khollam, Y. B.; Dhage, S. R.; Potdar, H. S.; Deshpande, S. B.; Bakare, P. P.; Kulkarni, S. D.; Date, S. K. Microwave Hydrothermal Preparation of Submicron-Sized Spherical Magnetite (Fe3O4) Powders. *Mater. Lett.* **2002,** *56* (4), 571–577.
37. Kim, J.; Kim, H.; Kim, J.; Lee, J. Ferrites: Emerging Light Absorbers for Solar Water Splitting. *J. Mater. Chem. A* **2020,** *8* (19), 9447–9482.
38. Korneva, G.; Ye, H.; Gogotsi, Y.; Halverson, D.; Friedman, G.; Bradley, J. C.; Kornev, K. G. Carbon Nanotubes Loaded with Magnetic Particles. *Nano Lett.* **2005,** *5* (5), 879–884.
39. Kukovitskii, E. F.; Chernozatonskii, L. A.; L'vov, S. G.; Mel'Nik, N. N. Carbon Nanotubes of Polyethylene. *Chem. Phys. Lett.* **1997,** *266* (3–4), 323–328.
40. Kumar, R.; Kumar, H.; Singh, R. R.; Barman, P. B. Variation in Magnetic and Structural Properties of Co-Doped Ni–Zn Ferrite Nanoparticles: A Different Aspect. *J. Sol-Gel Sci. Technol.* **2016,** *78* (3), 566–575.
41. Kyotani, T.; Tsai, L. F.; Tomita, A. Preparation of Ultrafine Carbon Tubes in Nanochannels of an Anodic Aluminum Oxide Film. *Chem. Mater.* **1996,** *8* (8), 2109–2113.
42. Laplaze, D.; Bernier, P.; Maser, W. K.; Flamant, G.; Guillard, T.; Loiseau, A. Carbon Nanotubes: The Solar Approach. *Carbon* **1998,** *36* (5–6), 685–688.
43. Lavela, P.; Tirado, J. L. CoFe2O4 and NiFe2O4 Synthesized by Sol–Gel Procedures for their Use as Anode Materials for Li Ion Batteries. *J. Power Sources* **2007,** *172* (1), 379–387.
44. Lee, J.; Ye, Y.; Ward, A. J.; Zhou, C.; Chen, V.; Minett, A. I.; Lee, S.; Liu, Z.; Chae, S. R.; Shi, J. High Flux and High Selectivity Carbon Nanotube Composite Membranes for Natural Organic Matter Removal. *Sep. Purif. Technol.* **2016,** *163*, 109–119.
45. Li, H.; Wang, J.; Huang, Y.; Yan, X.; Qi, J.; Liu, J.; Zhang, Y. Microwave Absorption Properties of Carbon Nanotubes and Tetrapod-Shaped Zno Nanostructures Composites. *Mater. Sci. Eng. B* **2010,** *175* (1), 81–85.
46. Lu, M.; Chang, Y.; Guan, X. H.; Wang, G. S. The Synthesis of Co x Ni 1− x Fe 2 O 4/Multi-Walled Carbon Nanotube Nanocomposites and their Photocatalytic Performance. *RSC Adv.* **2019,** *9* (58), 33806–33813.
47. Lu, Z.; He, M.; Yang, L.; Ma, Z.; Yang, L.; Wang, D.; Yan, Y.; Shi, W.; Liu, Y.; Hua, Z. Selective Photodegradation of 2-Mercaptobenzothiazole by a Novel Imprinted CoFe 2O4/MWCNTs Photocatalyst. *RSC Adv.* **2015,** *5* (59), 4782–47829.

48. Maaz, K.; Karim, S.; Mumtaz, A.; Hasanain, S. K.; Liu, J.; Duan, J. L. Synthesis and Magnetic Characterization of Nickel Ferrite Nanoparticles Prepared by Co-Precipitation Route. *J. Magn. Magn. Mater.* **2009,** *321* (12), 1838–1842.
49. Maaz, K.; Mumtaz, A.; Hasanain, S.; Ceylan, A. Synthesis and Magnetic Properties of Cobalt Ferrite (CoFe2O4) Nanoparticles Prepared by Wet Chemical Route. *J. Magn. Magn. Mater.* **2007,** *308* (2), 289–295.
50. Martinez-Vargas, S.; Martínez, A. I.; Hernández-Beteta, E. E.; Mijangos-Ricardez, O. F.; Vázquez-Hipólito, V.; Patiño-Carachure, C.; López-Luna, J. As (III) and As (V) Adsorption on Manganese Ferrite Nanoparticles. *J. Mol. Struct.* **2018,** *1154*, 524–534.
51. Mishra, N. R.; Mohanty, S. K.; Mittra, D.; Shah, M.; Meitei, W. B. Projecting Stunting and Wasting Under Alternative Scenarios in Odisha, India, 2015–2030: A Lives Saved Tool (LiST)-Based Approach. *BMJ Open* **2019**, *9* (5), e028681.
52. Murugesan, K.; Kalaichelvan, P. Synthetic Dye Decolourization by White Rot Fungi. *Indian J. Exp. Biol.* **2003,** *41*, 1076–1087.
53. Mustaffa, M. S.; Azis, R. S.; Abdullah, N. H.; Ismail, I.; Ibrahim, I. R. An Investigation of Microstructural, Magnetic and Microwave Absorption Properties of Multi-Walled Carbon Nanotubes/Ni 0.5 Zn 0.5 Fe 2 O 4. *Sci. Rep.* **2019;,** *9* (1), 1–7.
54. Nam, J. H.; Joo, Y. H.; Lee, J. H.; Chang, J. H.; Cho, J. H.; Chun, M. P.; Kim, B. I. Preparation of NiZn-Ferrite Nanofibers by Electrospinning for DNA Separation. *J. Magn. Magn. Mater.* **2009,** *321* (10), 1389–1392.
55. Naseri, M. G.; Saion, E. B.; Ahangar, H. A.; Hashim, M.; Shaari, A. H. Simple Preparation and Characterization of Nickel Ferrite Nanocrystals by a Thermal Treatment Method. *Powder Technol.* **2011,** *212* (1), 80–88.
56. Niu, W.; Pakhira, S.; Marcus, K.; Li, Z.; Mendoza-Cortes, J. L.; Yang, Y. Apically Dominant Mechanism for Improving Catalytic Activities of N-Doped Carbon Nanotube Arrays in Rechargeable Zinc–Air Battery. *Adv. Ener. Mater.* **2018,** *8* (20), 1800480.
57. Odling, G.; Robertson, N. Bridging the Gap Between Laboratory and Application in Photocatalytic Water Purification. *Catal. Sci. Technol.* **2019,** *9* (3), 533–545.
58. Odom, T.; Huang, J.; Kim, P.; Lieber, C. Atomic Structure and Electronic Properties of Single-Walled Carbon Nanotubes. *Nature* **1998,** *391* (6662), 62–64.
59. Oh, C.; Ji, S.; Cheong, Y.; Yim, G.; Applicability of Electrochemical Wastewater Treatment System Powered by Temperature Difference Energy. *J. Haz. Mater.* **2018,** *351*, 108–116.
60. Oyetade, O. A.; Nyamori, V. O.; Martincigh, B. S.; Jonnalagadda, S. B. Effectiveness of Carbon Nanotube–Cobalt Ferrite Nanocomposites for the Adsorption of Rhodamine B from Aqueous Solutions. *Rsc. Adv.* **2015,** *5* (29), 22724–22739.
61. Pan, B.; Xing, B. Adsorption Mechanisms of Organic Chemicals on Carbon Nanotubes. *Environ. Sci. Technol.* **2008,** *42* (24), 9005–9013.
62. Pare, B.; Sarwan, B.; Jonnalagadda, S. B. Photocatalytic Mineralization Study of Malachite Green on the Surface of Mn-Doped BiOCl Activated by Visible Light Under Ambient Condition. *Appl. Surf. Sci.* **2011,** *258* (1), 247–253.
63. Pham-Huu, C.; Keller, N.; Estournes, C.; Ehret, G.; Ledoux, M. J. Synthesis of CoFe 2 O 4 Nanowire in Carbon Nanotubes. A New Use of the Confinement Effect. *Chem. Commun.* **2002,** (17), 1882–1883.

64. Pham-Huu, C.; Keller, N.; Estournes, C.; Ehret, G.; Greneche, J. M.; Ledoux, M. J. Microstructural Investigation and Magnetic Properties of CoFe 2 O 4 Nanowires Synthesized Inside Carbon Nanotubes. *Phys. Chem. Chem. Phys.* **2003,** *5* (17), 3716–3723.
65. Prabhakaran, T.; Hemalatha, J. Combustion Synthesis and Characterization of Highly Crystalline Single Phase Nickel Ferrite Nanoparticles. *J. Alloys Comp.* **2011,** *509* (25), 7071–7077.
66. Priyadharsini, P. A. Pradeep, P. S. Rao and G. Chandrasekaran. *Mater. Chem. Phys*, **2009,** *116*, 207.
67. Pullar, R. C. Hexagonal Ferrites: A Review of the Synthesis, Properties and Applications of Hexaferrite Ceramics. *Prog. Mater. Sci.* **2012,** *57* (7), 1191–1334.
68. Raina, O.; Manimekalai, R. Photocatalysis of Cobalt Zinc Ferrite Nanorods under Solar Light. *Res. Chem. Intermed.* **2018,** *44* (10), 5941–5951.
69. Reddy, D. H. K.; Yun, Y. S. Spinel Ferrite Magnetic Adsorbents: Alternative Future Materials for Water Purification? *Coord. Chem. Rev.* **2016,** *315*, 90–111.
70. Ren, Z. F.; Huang, Z. P.; Xu, J. W.; Wang, J. H.; Bush, P.; Siegal, M. P.; Provencio, P. N. Synthesis of Large Arrays of Well-Aligned Carbon Nanotubes on Glass. *Science* **1998,** *282* (5391), 1105–1107.
71. Richter, H.; Hernadi, K.; Caudano, R.; Fonseca, A.; Migeon, H. N. Formation of Nanotubes in Low Pressure Hydrocarbon Flames. *Carbon (New York, NY)*, **1996,** *34*(3), 427–429.
72. Rondinone, A. J.; Samia, A. C.; Zhang, Z. J. Superparamagnetic Relaxation and Magnetic Anisotropy Energy Distribution in CoFe2O4 Spinel Ferrite Nanocrystallites. *J. Phys. Chem. B* **1999,** *103* (33), 6876–6880.
73. Salvetat, J.; Kulik, A.; Bonard, J.; Briggs, G.; Stöckli, T.; Méténier, K.; Bonnamy, S.; Béguin, F.; Burnham, N.; Forró, L.; Elastic Modulus of Ordered and Disordered Multiwalled Carbon Nanotubes. *Adv. Mat.* **1999,** *11* (2), 161–165.
74. Santa Maria, L. C.; Costa, M. A. S.; Soares, J. G. M.; Wang, S. H.; Silva, M. R. Preparation and Characterization of Manganese, Nickel and Cobalt Ferrites Submicron Particles in Sulfonated Crosslinked Networks. *Polymer* **2005,** *46* (25), 11288–11293.
75. See, C. H.; Harris, A. T. A Review of Carbon Nanotube Synthesis via Fluidized-Bed Chemical Vapor Deposition. *Indus. Eng. Chem. Res.* **2007,** *46* (4), 997–1012.
76. Singh, C.; Bansal, S.; Kumar, V.; Tikoo, K. B.; Singhal, S. Encrustation of Cobalt Doped Copper Ferrite Nanoparticles on Solid Scaffold CNTs and their Comparison with Corresponding Ferrite Nanoparticles: A Study of Structural, Optical, Magnetic and Photo Catalytic Properties. *RSC Adv.* **2015,** *5* (49), 39052–39061.
77. Singh, C.; Malik, R.; Kumar, V.; Singhal, S. CdS Nanorod–MFe 2 O 4 (M = Zn, Co and Ni) Nanocomposites: A Heterojunction Synthesis Strategy to Mitigate Environmental Deterioration. *RSC Adv.* **2015,** *5* (108), 89327–89337.
78. Singh, V. K.; Shukla, A.; Patra, M. K.; Saini, L.; Jani, R. K.; Vadera, S. R.; Kumar, N. Microwave Absorbing Properties of a Thermally Reduced Graphene Oxide/Nitrile Butadiene Rubber Composite. *Carbon* **2012,** *50* (6), 2202–2208.
79. Singh, C.; Bansal, S.; Kumar, V.; Singhal, S. Beading of Cobalt Substituted Nickel Ferrite Nanoparticles on the Surface of Carbon Nanotubes: A Study of their Synthesis

Mechanism, Structure, Magnetic, Optical and their Application as Photocatalyst. *Ceram. Int.* **2015,** *41* (3), 3595–3604.

80. Singh, C.; Bansal, S.; Singhal, S.; Synthesis of Zn1− xCoxFe2O4/MWCNTs Nanocomposites Using Reverse Micelle Method: Investigation of their Structural, Magnetic, Electrical, Optical and Photocatalytic Properties. *Physica B Condens. Matter*. **2014,** *444*, 70–76.
81. Shi, Y.; Ding, J.; Liu, X.; Wang, J. NiFe2O4 Ultrafine Particles Prepared by Co-Precipitation/Mechanical Alloying. *J. Magn. Magn. Mater.* **1999,** *205* (2–3), 249–254.
82. Kalikeri, S.; Kodialbail, V. S. *Environ. Sci. Pollut. Res.* **2018,** *25*, 1–13.
83. Smedley, P. Andkinniburgh, D. G. Behaviour and Distribution of Arsenic in Natural Waters. *Appl. Geochem.* **2002,** *17*, 517–568.
84. Sontu, U. B.; Chou, F. C. Temperature Dependent and Applied Field Strength Dependent Magnetic Study of Cobalt Nickel Ferrite Nano Particles: Synthesized by an Environmentally Benign Method. *J. Magn. Magn. Mater.* **2018,** *452*, 398–406.
85. Sunny, V.; Kurian, P.; Mohanan, P.; Joy, P. A.; Anantharaman, M. R. A Flexible Microwave Absorber based on Nickel Ferrite Nanocomposite. *J. Alloys Comp.* **2010,** *489* (1), 297–303.
86. Tan, L.; Shuang, C.; Wang, Y.; Wang, J.; Su, Y.; Li, A. Effect of Pore Structure on the Removal of Clofibric Acid by Magnetic Anion Exchange Resin. *Chemosphere* **2018,** *191*, 817–824.
87. Tan, T.; Xie, W.; Zhu, G.; Shan, J.; Xu, P.; Li, L.; Wang, J. Fabrication and Photocatalysis of BiFeO 3 with Inverse Opal Structure. *J. Porous Mater.* **2015,** *22* (3), 659–663.
88. Teo, K. B.; Singh, C.; Chhowalla, M.; Milne, W. I. Catalytic Synthesis of Carbon Nanotubes and Nanofibers. *Encyclo. Nanosci. Nanotechnol.* **2003,** *10* (1).
89. Terrones, M.; Hsu, W. K.; Terrones, H.; Zhang, J. P.; Ramos, S.; Hare, J. P.; Castillo, R.; Prasides, K.; Cheetham, A. K. HW Kroto, and DRM Walton. *Chem. Phys. Lett,* **1996,** *259*, 568.
90. Thess, A.; Lee, R.; Nikolaev, P.; Dai, H.; Petit, P.; Robert, J.; Xu, C.; Lee, Y. H.; Kim, S. G.; Rinzler, A. G.; Colbert, D. T. Crystalline Ropes of Metallic Carbon Nanotubes. *Science* **1996,** *273* (5274), 483–487.
91. Tolani, S. C.; Golhar, A. R.; Rewatkar, K. G. In *A Review of Morphological, Structural Behaviour and Technological Applications of Ferrites*, AIP Conference Proceedings, Vol. 2104, No. 1, P. 030032); AIP Publishing LLC, 2019.
92. Tyagi, S.; Baskey, H. B.; Agarwala, R. C.; Agarwala, V.; Shami, T. C. Development of Hard/Soft Ferrite Nanocomposite for Enhanced Microwave Absorption. *Ceramics Int.* **2011,** *37* (7), 2631–2641.
93. Tyagi, S.; Pandey, V. S.; Baskey, H. B.; Tyagi, N.; Garg, A.; Goel, S.; Shami, T. C. RADAR Absorption Study of BaFe12O19/ZnFe2O4/CNTs Nanocomposite. *J. Alloys Comp.* **, 2018,** *731*, 584–590.
94. Vander Wal, R. L.; Berger, G. M.; Hall, L. J. Single-Walled Carbon Nanotube Synthesis via a Multi-Stage Flame Configuration. *J. Phys. Chem. B*, **2002,** *106* (14), 3564–3567.

95. Vander Wal, R. L.; Ticich, T. M.; Flame and Furnace Synthesis of Single-Walled and Multi-Walled Carbon Nanotubes and Nanofibers. *J. Phys. Chem. B*, **2001,** *105* (42), 10249–10256.
96. Verma, A.; Dube, D. C. Processing of Nickel–Zinc Ferrites via the Citrate Precursor Route for High-Frequency Applications. *J. Am. Ceramic Soc.* **2005,** *88* (3), 519–523.
97. Wadhawan, A.; Garrett, D.; Pérez, J. M. Nanoparticle-Assisted Microwave Absorption by Single-Wall Carbon Nanotubes. *Appl. Phys. Lett.* **2003,** *83* (13), 2683–2685.
98. Wang, J. Prepare Highly Crystalline NiFe2O4 Nanoparticles with Improved Magnetic Properties. *Mater. Sci. Eng. B* **2006,** *127* (1), 81–84.
99. Wang, L.; Xing, H.; Gao, S.; Ji, X.; Shen, Z. Porous Flower-Like NiO@ Graphene Composites with Superior Microwave Absorption Properties. *J. Mater. Chemi. C* **2017,** *5* (8), 2005–2014.
100. Wang, Q.; Yang, J.; Dong, Y.; Zhang, L. One-Step Fabrication of a Multifunctional Magnetic Nickel Ferrite/Multi-Walled Carbon Nanotubes Nanohybrid-Modified Electrode for the Determination of Benomyl in Food. *J. Agric. Food Chem.* **2015,** *63* (19), 4746–4753.
101. Wang, Z.; Liu, X.; Lv, M.; Chai, P.; Liu, Y.; Meng, J. Preparation of Ferrite MFe2O4 (M = Co, Ni) Ribbons with Nanoporous Structure and their Magnetic Properties. *J. Phys. Chem. B* **2008,** *112* (36), 11292–11297.
102. Wepasnick, K. A.; Smith, B. A.; Schrote, K. E.; Wilson, H. K.; Diegelmann, S. R.; Fairbrother, D. H. Surface and Structural Characterization of Multi-Walled Carbon Nanotubes Following Different Oxidative Treatments. *Carbon*, **2011,** *49* (1), 24–36.
103. Wilder, J.; Venema, L.; Rinzler, A.; Smalley, R.; Dekker, C. Electronic Structure of Atomically Resolved Carbon Nanotubes. *Nature* **1998,** *391* (6662), 59–62.
104. Wojnarovits, L.; Takacs, E. Irradiation Treatment of Azo Dye Containing Wastewater: An Overview. *Radiat. Phys. Chem.* **2008,** *77* (3), 225–244.
105. Wu, M.; Lin, G.; Chen, D.; Wang, G.; He, D.; Feng, S.; Xu, R. Sol-Hydrothermal Synthesis and Hydrothermally Structural Evolution of Nanocrystal Titanium Dioxide. *Chem. Mater.* **2002,** *14* (5), 1974–1980.
106. Wu, Z.; Zhang, R.; Yu, Z.; Shan, L.; Dong, L.; Zhang, X. Study on Preparation and Magnetic Properties of Sr1-x Gd x Fe12-x Cu x O19 ($0.00 \leq x \leq 0.20$) Strontium Ferrite Prepared by Solid Phase Method. *Ferroelectrics* **2018,** *523* (1), 82–88.
107. Yamamoto, K.; Koga, Y.; Fujiwara, S.; Kubota, M. New Method of Carbon Nanotube Growth by Ion Beam Irradiation. *Appl. Phys. Lett.* **1996,** *69* (27), 4174–4175.
108. Yan, Y.; Miao, J.; Yang, Z.; Xiao, F. X.; Yang, H. B.; Liu, B.; Yang, Y. Carbon Nanotube Catalysts: Recent Advances in Synthesis, Characterization and Applications. *Chem. Soc. Rev.* **2015,** *44* (10), 3295–3346.
109. Yan, Y.; Xu, Y.; Zhao, B.; Xu, Y.; Gao, Y.; Chen, G.; Wang, W.; Xia, B. Y. Bifunctional Nickel Ferrite-Decorated Carbon Nanotube Arrays as Free-Standing Air Electrode for Rechargeable Zn–Air Batteries. *J. Mater. Chem. A* **2020,** *8* (10), 5070–5077.
110. Zahrim, A. Y.; Hilal, N. Treatment of Highly Concentrated Dye Solution by Coagulation/Flocculation–Sand Filtration and Nanofiltration. *Water Res. Indus.* **2013,** *3*, 23–34.
111. Zhang, D.; Pan, B.; Cook, R. L.; Xing, B. Multi-Walled Carbon Nanotube Dispersion by the Adsorbed Humic Acids with Different Chemical Structures. *Environ. Pollut.* **2015,** *196*, 292–299.

112. Zhang, J.; Shi, J.; Gong, M. Synthesis of Magnetic Nickel Spinel Ferrite Nanospheres by a Reverse Emulsion-Assisted Hydrothermal Process. *J. Solid State Chem.* **2009,** *182* (8), 2135–2140.
113. Zhang, Y. M.; Xu, P. L.; Zeng, Q.; Liu, Y. M.; Liao, X.; Hou, M. F. Magnetism-Assisted Modification of Screen Printed Electrode with Magnetic Multi-Walled Carbon Nanotubes for Electrochemical Determination of Dopamine. *Mater. Sci. Eng. C* **2017,** *74*, 62–69.
114. Zhang, Z.; O'Hara, I.; Kent, G.; Doherty, W. Comparative Study on Adsorption of Two Cationic Dyes by Milled Sugarcane Bagasse. *Indus. Crops Prod.* **2013,** *42*, 41–49.
115. Zhao, T.; Ji, X.; Jin, W.; Wang, C.; Ma, W.; Gao, J.; Dang, A.; Li, T.; Shang, S.; Zhou, Z. Direct In Situ Synthesis of A 3D Interlinked Amorphous Carbon Nanotube/Graphene/BaFe 12 O 19 Composite and Its Electromagnetic Wave Absorbing Properties. *RSC Adv.* **2017,** *7* (26), 15903–15910.
116. Zhao, X.; Fu, Y.; Wang, J.; Xu, Y.; Tian, J. H.; Yang, R. Ni-Doped CoFe2O4 Hollow Nanospheres as Efficient Bi-Functional Catalysts. *Electrochimica Acta* **2016,** *201*, 172–178.
117. Zheng, Z.; Xu, B.; Huang, L.; He, L.; Ni, X. Novel Composite of Co/Carbon Nanotubes: Synthesis, Magnetism and Microwave Absorption Properties. *Solid State Sci.* **2008,** *10* (3), 316–320.
118. Zhou, J.; Ma, J.; Sun, C.; Xie, L.; Zhao, Z.; Tian, H.; Wang, Y.; Tao, J.; Zhu, X. Low Temperature Synthesis of NiFe2O4 by a Hydrothermal Method. *J. Am. Ceramic Soc.* **2005,** *88* (12), 3535–3537.
119. Zhou, X.; Shen, L.; Li, L.; Zhou, S.; Huang, T.; Hu, C.; Pan, W.; Jing, X.; Sun, J.; Gao, L.; Huang, Q. Microwave Sintering Carbon Nanotube/NiO. 5ZnO. 5Fe2O4 Composites and their Electromagnetic Performance. *J. Eur. Ceramic Soc.* **2013,** *33* (11), 2119–2126.

CHAPTER 4

HETEROATOM-DOPED GRAPHENE FOR ENERGY STORAGE APPLICATIONS

JAISON M. JOY[1], MOHAMMAD REZA SAEB[2], JYOTISHKUMAR PARAMESWARANPILLAI[3], and C. D. MIDHUN DOMINIC[1,*]

[1]*Department of Chemistry, Sacred Heart College (Autonomous), Kochi, Kerala 682013, India*

[2]*Department of Polymer Technology, Faculty of Chemistry, Gdańsk University of Technology, Gabriela Narutowicza 11/12, 80-233 Gdańsk, Poland*

[3]*School of Biosciences, Mar Athanasios College for Advanced Studies Tiruvalla (MACFAST), Pathanamthitta, Kerala 689101, India*

**Corresponding author. E-mail: midhundominic@shcollege.ac.in*

ABSTRACT

In line with the upward demand for renewable energy resources, researchers have made tremendous efforts to develop sustainable graphene-based energy storage systems. Heteroatom doping of graphene significantly boosts the performance of emerging technologies like electrochemical energy storage systems. Various approaches have been used to achieve heteroatom doping in graphene for tunable chemical, electrical, and mechanical properties. Benefiting from such a distinct property,

Nanostructured Carbon for Energy Generation, Storage, and Conversion. V. I. Kodolov, DSc, Omari Mukbaniani, DSc, Ann Rose Abraham, PhD, and A. K. Haghi, PhD (Eds.)

heteroatom-doped graphene can provide a key to the development of novel materials for effective energy storage applications. In this view, we summarize and discuss the latest progress in the synthesis and applications of heteroatom-doped graphene in batteries and supercapacitors.

4.1 INTRODUCTION

Energy production and storage have incurred intense research interest arising from the need for highly efficient energy resources. A greater focus is given to the area of green energy production and consumption. Renewable energy sources such as wind, tidal, and solar energies are accounted for clean electricity generation. However, to implement extensive usage of renewable energy sources, material abundance, easy, and clean technologies are required. Moreover, efficient energy storage devices should be developed to meet the world's electricity needs.[1]

Carbon materials have attracted strong interest as electrode materials for electrochemical energy storage systems. The excellent thermo-mechanical and physicochemical properties of *carbon*-based *materials* give rise to explore their *applications* in the field of *energy storage.*[2] Recently, graphene materials are being examined and extensively developed for several applications due to their unique properties, such as high thermal and electrical conductivity, large surface area, high strength, and chemical resistance.[3,4] To improve the properties of graphene, as an excellent energy storage material, great efforts have also been spent, mainly by the modification of graphene with polymers, metal nanoparticles, and doping with heteroatoms.[5] However, currently, heteroatom-doped graphene has been focused on as an ideal electrode material with inherent functionality in energy storage applications.[6] In this chapter, recent developments in this field including various heteroatom doped graphene materials and their usage in energy-related technologies such as batteries and supercapacitors were discussed.

4.2 HETEROATOM DOPING IN GRAPHENE

Heteroatom doping of graphitic material can alter their electronic and chemical properties, making them suitable for the production of energy storage devices. The doping process can be performed in two ways. In the first method, carbon atoms from the graphene network are replaced by heteroatoms, while in the second method, the dopant is adsorbed onto the graphene surface.[7] The first way of doping creates sp^3 defects in the graphene through covalent bonding, whereas in the second method, dopant adsorption on the surface of graphene does not trigger any defect. Multiple or co-doping of heteroatoms can deliver a synergistic effect of both the dopants. On the other hand, the nature of the dopant and the percentage of doping can alter the intrinsic properties of graphene.[8]

Heteroatom doping in the graphitic type of materials like graphene nanosheets (GNS), graphene nano-ribbons, graphene quantum dots, graphene oxide (GO), and reduced graphene oxide can effectively improve its physico-chemical properties.[9] Introduction of heteroatoms (boron, nitrogen, phosphorus, sulfur, Iodine, bromine, chlorine, and fluorine) in the graphene network allows the modification of properties such as, charge transport, bandgap, Fermi level, electrical properties, optical properties, magnetic properties, and thermal properties[10] Figure 4.1 reflects the schematic representation of various heteroatom doping in graphene sheet.[11] Due to the similarities in properties with the carbon atoms, these heteroatoms are suitable candidates as dopants in the graphitic network.

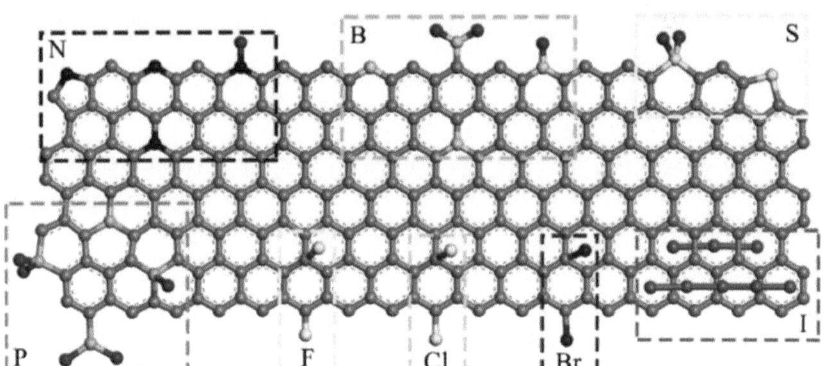

FIGURE 4.1 Heteroatom doping (N, B, S, P, F, Cl, Br, I) in a single graphene sheet.[11]
Source: Reprinted with permission from Ref. [11]. © 2019 Elsevier.

Nitrogen bonding with carbon creates an almost similar bond length as that of C–C. Nitrogen doping in graphene can form three different configurations; namely graphitic (or quaternary), pyridinic, and pyrrolic types. Pyrrolic type of nitrogen bonded with sp^3 configuration disrupts the planar structure of graphene, conversely, no considerable effect is detected from the pyridinic and graphitic type of doping.[12,13] Nitrogen is more electronegative than carbon; thus, a polarization between these two elements influence the alteration in properties of graphene in the course of doping.[14] Boron is an element with one less valence electron with respect to carbon, and it can form sp^2 hybridization in the graphene lattice. This helps to retain the planar structure of the graphene, but the charge polarization between the carbon and boron may affect some other properties.[15] In the case of phosphorous doping, P–C bond length formed is more than C–C bond length, and phosphorous shows low electronegativity (2.19) than the carbon, making more structural distortion in the graphene. Phosphorous forms a pyramidal structure with the carbon network and converts sp^2 hybridized carbon into sp^3 state.[16] Sulfur doping in the graphene also causes structural disruption because of the difference in carbon-sulfur (1.78 Å) and carbon–carbon (1.42 Å) bond lengths. However, sulfur forms similar functional groups like oxygen on doping in the carbon network.[17] Moreover, there is only a small polarization existing between carbon–sulfur bond due to their similar electronegativity. More reactive halogens (F, Cl, Br, I) can also change the sp^2 carbon to the sp^3 state on doping with graphene. As a result of halogen doping, graphene undergoes some changes in its geometric and electronic properties.

4.3 SYNTHESIS OF HETEROATOM-DOPED GRAPHENE

There are a number of different synthesis methods have been developed for the doping of heteroatom in graphene materials. The heteroatom doping can be categorized into single-step and multiple-step synthesis approaches. In a single-step approach, one can attain graphene synthesis and heteroatom doping simultaneously in a one-pot process. Alternatively, in multiple-step synthesis, the graphene or GO has to be prepared initially then heteroatom doping process can be followed in the next step. For example, Lin et al.[18] prepared boron-doped graphene from carbon tetrachloride, potassium, and tribromoboron using a single-step solvothermal process, while, in the work of Thirumal et al.[19] GO was first prepared using modified Hummer's method, and then GO is thermally reduced to GNS. Finally, boron was doped in reduced GNS using H_3BO_3 by hydrothermal method.

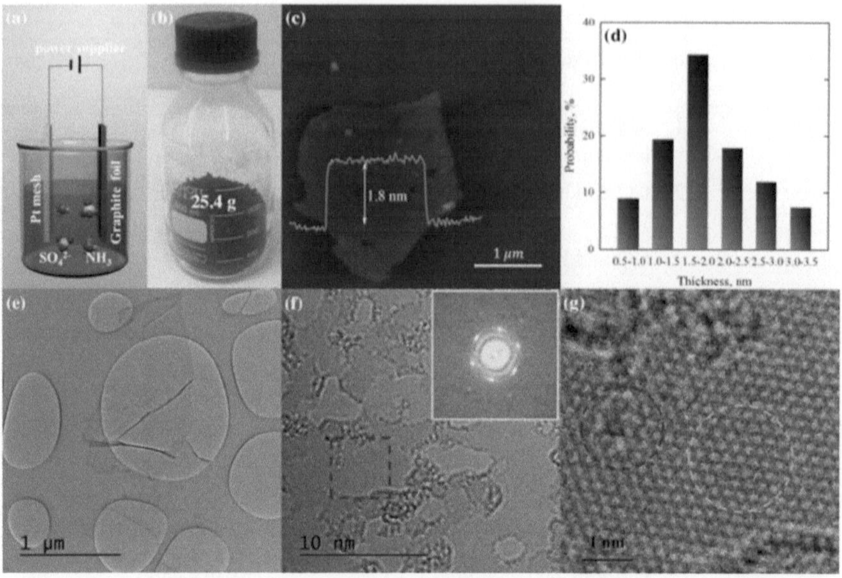

FIGURE 4.2 (a) Schematic of the preparation of nitrogen-doped graphene. (b) Photograph of nitrogen doped graphene powder. (c) AFM image of nitrogen-doped graphene with a height profile (blue curve) taken along the red line (d) The average thickness distribution of nitrogen-doped graphene powder obtained from the AFM height image. (e) TEM image of nitrogen-doped graphene powder. (f) High-magnification TEM image of nitrogen-doped graphene powder. (Inset is the FFT of the image). (g) The red box in figure F is enlarged.[21]

Source: Reprinted with permission from Ref. [21]. © 2012 Royal Society of Chemistry.

Various other doping strategies include chemical vapor deposition (CVD), ball milling, thermal annealing, plasma and arc discharge, etc. Several chemicals have been used as a precursor for heteroatom doping such as urea or ammonia as nitrogen source, H_3BO_3 or diborane as boron source, and carbon disulfide or hydrogen sulfide as sulfur source. The doping percentage can be varied by using different precursors or synthesis methods. For instance, Deng et al.[20] reported the one-pot solvothermal synthesis of nitrogen-doped graphene using tetrachloromethane with lithium nitride. This method induced 4–16% of nitrogen content in the graphitic structure. In another study, nitrogen-doped graphene was prepared by electrochemical exfoliation of graphite in an aqueous electrolyte containing ammonium sulfate and ammonium hydroxide (Fig. 4.2). Using this method, nitrogen doping present in the graphene was found to be 4.95%.[21] In Table 4.1, methods for the synthesis of heteroatom doped graphene are discussed in detail.

TABLE 4.1 Methods of Synthesis of Heteroatom-Doped Graphene.

Doping element	No	Precursors	Percentage of doping	Methods	Reference
Nitrogen	1	Methane/Ni on SiO_2/Si and ammonia	N/C atomic ratio 4%	CVD	22
	2	GO and ammonia	5.06%	Thermal annealing	23
	3	GO and Polypyrrole	2–3%	Pyrolysis	24
	4	GO and nitrogen gas	1.68–2.51%	Plasma process	25
	5	GO and Ammonium hydroxide	7.2%	Hydrothermal	26
	6	Lithium nitride and tetrachloromethane	10.5%	Solvothermal	27
	7	Graphite oxide and ammonia	5.04 wt%	Microwave heating	28
	8	Carbon nanotube	4%	Electrochemical	29
	9	Graphite and ammonia	1%	Arc discharge	30
Boron	10	Boron and Ethanol	0.5%	CVD	31
	11	Boric acid and graphite oxide	5.93%	Thermal reduction	32
	12	Graphite and boron	2–3%	Pulsed laser vaporization	33
	13	Graphite oxide and BH_3	1.1%	Liquid reflux process	34
	14	Graphite and boron/diborane	3.1%	Arc discharge	35
	15	GO and BCl_3	0.88%	Thermal reduction	36
	16	CCl_4, potassium and BBr_3	2.56%	reductive coupling process	37
Sulfur	17	Hexane and Sulfur	~0.6%	CVD	38
	18	Graphite and sulfur	4.94%	Ball milling process	39
	19	Graphite oxide and H_2S	1.2–1.7%	Thermal annealing	40
	20	Graphite oxide and H_2S	11.99 wt.%	Thermal exfoliation process	41

TABLE 4.1 *(Continued)*

Doping element	No	Precursors	Percentage of doping	Methods	Reference
	21	Graphene and CS_2	2.3%	Plasma treatment	42
Phosphorus	22	GO and 1-butyl-3-methlyimidazolium hexafluorophosphate	1.16%	Thermal annealing	43
	23	CH_4 and $Mg_3(PO_4)_2$	0.72%	CVD	44
	24	Graphite and H_3PO_4	2.4%	Hydrothermal/ annealing	45
	25	Graphite oxide and triphenylphosphine	1.32%	Thermal annealing	46
Iodine	26	GO and iodine	1.21 wt.%	Thermal annealing	47
	27	Graphite oxide and iodine	0.06 wt.%	Pyrolysis	48
	28	GO and potassium iodide	0.52 wt.%	Hydrothermal	49
Bromine	29	Graphene and Bromine vapor	0.4 wt.%	Room temperature addition reaction	50
	30	1,2,5,6,9,10-Hexabromo-cyclododecane	3.5 wt.%	CVD	51
Chlorine	31	Graphite and perchloric acid	3.45 wt%	Electrochemical	52
	32	GO and hydrochloric acid	1.01 wt.%	Reduction process	53
Fluorine	33	GO and zinc fluoride	2.61 wt.%	Thermal pyrolysis	54
	34	Graphite fluoride	10 wt.%	Arc-discharge process	55

Graphene containing two or more heteroatom at different composition could result in material with significant electronic structures and properties. Co-doping or multiple doping of graphene with different elements has been reported. Table 4.2 summarizes the various methods of preparing co-doped graphene.

TABLE 4.2 Synthesis Methods of Heteroatom Co-doping in Graphene.

	Precursor	Method	Doping elements and percentage of doping	Reference
1	GO and ammonia boron trifluoride	Hydrothermal	Boron – 0.6% and Nitrogen – 3.0%	56
2	GO and L-cysteine	Hydrothermal	Nitrogen – 0.9% and Sulfur – 0.3%	57
3	Graphite oxide, triphenylphosphine, and ammonia	Thermal annealing and hydrothermal reaction	Nitrogen – 4.38% and Phosphorous – 1.93%	58
4	GO, thioglycolic acid and phytic acid	Chemical reduction	Sulfur – 5.8% and phosphorous – 4.6%	59
5	GO and trimethylamine tri(hydrofluoride)	Hydrothermal	Nitrogen – 3.24% and fluorine – 10.9%	60
6	GO, boric acid, cyanamide, and phenylphosphine	Pyrolysis	Nitrogen – 13.12%, boron – 17.52 %, and Phosphorous – 0.43%	61
7	GO and ammonium hexafluorophosphate	Thermal annealing	Nitrogen – 7.11% and phosphorous – 0.37%, Fluorine – 0.33%	62

4.4 IMPORTANCE OF HETEROATOM DOPING IN GRAPHENE

Heteroatom doping causes modifications in the electronic properties of graphene. The perfect hexagonal honeycomb lattice structure of graphene is possibly altered by the inclusion of heteroatoms. These structural distortions can lead to modification of graphene properties including drastic changes in bandgap, Fermi level, charge transport, and spin density. Moreover, it can even affect thermal and chemical stability, optical characteristics, and magnetic properties.[63] However, improved or modified properties may depend upon several factors such as type and percentage of dopants, reaction conditions, etc. For instance, heteroatom doping can increase the wettability of graphene, which is crucial for supercapacitor.[64] The introduction of heteroatom to the basal planes of graphene sheet can create active sites, increase its reactivity and electronic conductivity. Moreover, the presence of co-dopants in graphene structure causes a synergistic effect, which in turn results in additional electrochemical properties.[65]

Thus, it is important to understand how to engineer the graphene properties by heteroatom doping for a particular application.

Nitrogen doping generates polarization with carbon in the graphene network, which influences the electronic state, changing zero bandgap metal state to semiconducting properties. A graphitic nitrogen doping contributes electron toward the lattice structure, giving an n-type doping effect. On the other hand, pyridinic and pyrrolic type of nitrogen doping creates a p-type doping effect since it consumes electron from the graphene lattice. Studies showed that graphitic nitrogen doping (0.4 at.%) induces 0.3 eV increment in the bandgap and charge-carrier concentration of 8×10^{12} cm^{-2}.[12] Ouerghi et al.[66] reported a large carrier concentration of 2.6×10^{13} cm^{-2} for 0.6 at% graphitic nitrogen doping, although, presence of pyridinic and pyrrolic nitrogen creates no change.

Moreover, pyrrolic nitrogen doping can cause spin polarization due to the presence of non-bonding electrons in π and π^* states. Thus, the pyrrolic type of nitrogen doping generates a strong magnet moment and pyridinic type has a weak effect. However, the graphitic nitrogen type of doping creates no magnetic moment due to the absence of nonbonding electrons.[67] Furthermore, all types of nitrogen doping result in changes in the optical properties of graphene as it cause a blue shift in the photoluminescence peak, while graphitic nitrogen doping can increase the intensity of the peak.[68]

Boron doping generates a P-type doping due to electron-deficient nature of boron, which results in down shift in the Fermi level and opening of bandgap. It is evident that the bandgap mostly depends upon number of graphene layers and boron doping concentration. The Fermi level decreases by 0.65 eV on doping 2 wt.% of in-plane boron in graphene.[15,69] Boron doping in graphene distorts the planar structure although it brings excellent mechanical properties due to strong B–C bond. In plane boron doping in graphene is easier than nitrogen doping as boron forms sp^2 hybridization in the carbon lattices. However, boron doping causes reduction in the thermal conductivity of graphene.[70]

Phosphorous doping delivers a n-type behavior to the graphene and electron mobility is 5 times higher than the pristine bilayer graphene. The studies showed that bandgap opening directly depends on phosphorous doping and at 0.5% of doping in graphene can results in a bandgap of 0.3–0.4 eV. Phosphorous doped graphene also shows magnetic moment due to distortion in the symmetry of graphene π-electron framework.[71]

Sulfur doping provides a small-band-gap semiconductor or more metallic than pristine graphene. The resistivity of sulfur-doped graphene increases as it traps the free charge carriers by forming functional groups. Sulfur in the carbon network creates a nonuniform spin density distribution, which eventually offers catalytic properties toward applications like oxygen reduction reactions.[72]

Halogen atom doping in graphene lattice results in alteration of geometric and electronic properties. Fluorine-doped graphene has attained a high mechanical strength and chemical inertness. Fluorine doping in graphene increases hydrophobicity. Studies showed that fluoro-graphene is the thinnest insulator since it has a wide bandgap of ~3 eV. In addition, these bandgap can be tunable by the amount of fluorine coverage on graphene.[73] The introduction of chlorine to graphene generates a p-type doping. Covalently bonded chlorine at 25% in graphene produces a bandgap of 1.4 eV. This creates a high hole concentration of about 1.2×10^{13} cm^2. Moreover, chlorine doping gives high carrier mobility of 1535 cm^2 V^{-1} S^{-1} and shows an increment in the conductivity by 2 times than the pristine graphene.[74–76] There are a few reports on bromine and Iodine doping on graphene. These large-sized halogen atoms can form a link with graphene through physisorption or charge transfer complex formation.[77] Co-doped or multidoped graphene generates the properties of both the dopants present in it, i.e., it will create a synergistic effect. For instance, boron and nitrogen co-doped graphene shows thermal stability higher than boron-doped graphene and lower than that of nitrogen-doped graphene. Co-dopant can change the bonding configuration of other dopants in the graphene. Mostly pyrrolic type of nitrogen doping is found higher in nitrogen-doped graphene, while the graphitic type of nitrogen is more in sulfur co-doped graphene.[78,79]

4.5 HETEROATOM DOPED GRAPHENE IN ENERGY STORAGE APPLICATIONS

Doping of various heteroatom into graphene networks offer intriguing properties, which are excellent for some potential applications. Thus, doped graphene can be useful in a variety of fields such as medical and electronic applications. These heteroatom-doped graphene have been also used in promising fields, including field-effect transistors, photoluminescence, light-emitting diodes, electrochemical bio-sensing, and energy storage applications.[7] However, in particular, we discuss the application of hetero-atom doped graphene for electrochemical energy storage devices like batteries and supercapacitors.

4.6 SUPERCAPACITORS

Supercapacitors, also known as high-capacity capacitors, because of their high surface area and thin electrolytic dielectrics allow gaining more energy than conventional capacitors. It shows higher specific power, short charging times, and long life than batteries. In comparison with other energy storage devices, it has simple electrical components, longer lifetime, and shows no memory effect. Consequently, supercapacitors are shown to be an outstanding competent for sustainable energy storage applications. On the other hand, the energy density associated with the supercapacitors was significantly lower than batteries.[80] That is, a battery can store more energy in a specific volume than a supercapacitor, which means its energy density is high.

Supercapacitors operate on the same principle as conventional capacitors. Dielectric plates are used in conventional capacitors for electrostatic charge storage. Supercapacitor consists of porous electrodes in an electrolyte solution, and kept a separator to hinder the direct contact between the electrodes, whereas it helps in movement of charged ions. The supercapacitors are categorized into three types according to their storage mechanism, namely, Electrical Double Layer Capacitor (EDLC), pseudo capacitor, and hybrid supercapacitor. In EDLC, energy storage mechanism is based on intrinsic electrode area and atomic charge partition length, while energy storage in pseudo capacitor is achieved by reversible redox reactions among electro-active materials on electrode and electrolyte solution. The hybrid supercapacitors configuration results from the combination of redox and EDLC materials as electrode to provide high energy density.[81] Supercapacitors possess higher surface area electrodes, which give larger capacitance value by a factor of 10^4 than conventional capacitors.[82] Incorporating electrodes with higher surface areas and keeping a very small distance between the electrodes can increase the capacitance. These electrochemical storage devices work on the mechanism of adsorption of ions from electrolytes onto the electrodes surface area.

The graphene-based materials are a good choice for supercapacitors due to their high surface area, and good electrical conductivity. However, graphene has a bandgap of zero, which limits its application as an active material in supercapacitors. Bandgap can be altered using several approaches like modification or chemical doping. Thus heteroatom doping is an efficient method to alter the electrical properties of graphene.[83] Such chemically doped graphene has shown excellent supercapacitor properties, for example, nitrogen-doped graphene exhibited a specific capacitance of 461 F g^{-1} at a scan rate of 5 mV s^{-1}.[84] In another study, boron-doped graphene, prepared from boric acid and GO, showed a specific capacitance of 172.5 F g^{-1}. On comparison with pristine graphene, it showed a 80% of increment in the capacitance mostly due the presence of boron-oxygen functional groups (Fig. 4.3).[85]

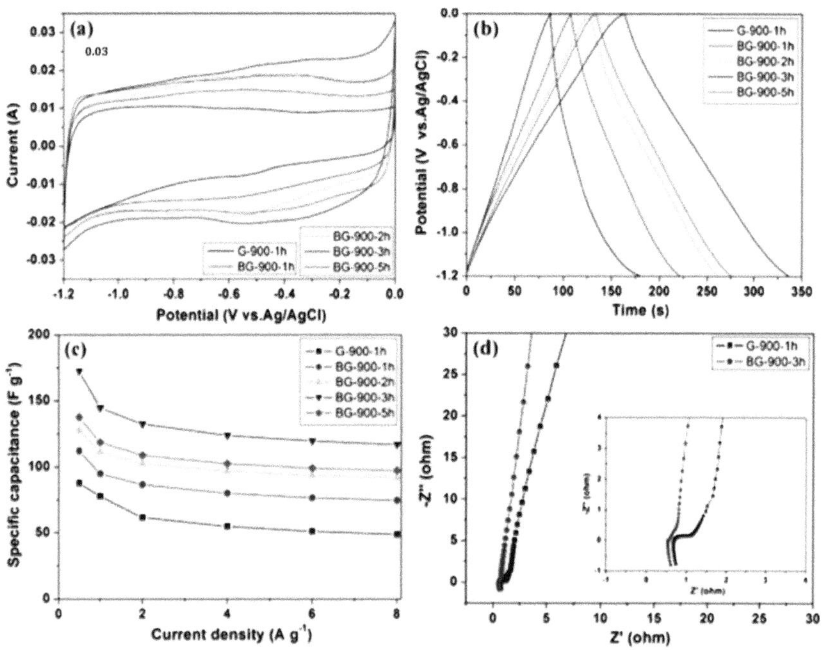

FIGURE 4.3 Current versus potential curves measured at the scan rate of 40 mVs^{-1} (a), charge–discharge curves (b), specific capacitance versus current density of graphene and BG from 0.5 to 8 Ag^{-1} (c), and electrochemical Impedance spectra of graphene and BG-900-3h (d).

Source: Reprinted with permission from Ref. [85]. © 2013 Elsevier.

Cheng et al.[86] developed co-doped graphene-based materials such as nitrogen-sulfur (N, S) co-doped graphene for supercapacitor applications. In this study, N, S co-doped graphene exhibited a charge-discharge cycling stability of up to 25000 cycles. Enhancement in the specific capacitance is mostly attributed to the strong irradiation used for the preparation, increased surface area, and pyrrolic nitrogen content. However, both doped and co-doped graphene materials have been used as supercapacitor electrodes. In Table 4.3, various hetero-atom doped/co-doped graphene used for supercapacitor applications are listed.

TABLE 4.3 Various Heteroatom-Doped Graphene and their Super Capacitive Properties.

Doped material	Specific capacitance of	Advantages	Ref
Nitrogen doped graphene	381 F g^{-1} at a current density of 1 A g^{-1}	High electron mobility, larger space charge capacitance.	87
Boron doped graphene	90 F g^{-1} at a current density of 2 A g^{-1}	Superior electrochemical performance is due to the p-type or hole conductivity	88
Phosphorus-doped graphene	115 F g^{-1} at a current density of 0.05 A g^{-1}	Cycling stability, stable at a wide voltage window	89
Sulfur-doped graphene	320 F g^{-1} at a current density of 3 A g^{-1}	Excellent cycling stability and efficiency	90
Fluorine-doped graphene	279.8 F g^{-1} at a current density of 0.5 A g^{-1}	high electronic conductivity and electro-catalytic active sites	91
Nitrogen, boron co-doped graphene	132 F g^{-1}at a scan rate of 100 mV s^{-1}	High surface area, 3D macroporosity, and high electrical conductivity	56
Nitrogen, sulfur co-doped graphene	431 F g^{-1} at a current density of 0.5 A g^{-1}	Free-standing electrode in supercapacitor, good mechanical properties	92
Sulfur and phosphorus co-doped graphene	438 F/g^{-1} at a scan rate of 10 mV/s	electrochemical stability, capacitance retention	59

4.7 BATTERIES

Rechargeable batteries are used in powering portable electronics, power tools, electric vehicles, etc. Among rechargeable batteries, lithium–ion batteries (LIBs) are most prominent, however, a variety of other battery types such as lithium–sulfur batteries (LSBs), sodium-ion batteries (SIBs), potassium ion batteries (KIBs), and metal-air batteries are also popular nowadays. Batteries have high specific energy density than supercapacitors. LIBs are the most employed battery type, due to its high cycling stability and reversibility. LIBs consist of a cathode (transition metal oxide), anode (graphite), a polymer porous membrane separator, and an organic electrolyte. The energy storage mechanism was explained by lithium-ion movement from anode to cathode during charging and cathode to anode during the discharging process. Same kind of intercalation and de-intercalation process has been followed by sodium and potassium ions in SIBs and KIBs, respectively. Thus, graphite anode plays an important role in these battery performances. However, graphene materials work in a different way in LSBs and metal-air batteries. Mostly graphitic materials provide a framework with high electron conductivity, rapid ionic movement, and stability to the electrodes.[93]

In LIBs, graphene shows a high reversible capacity due to large surface area, high conductivity, and stability than commercial graphite anode.[94] It is expected that the heteroatom-doped graphene can show better electrochemical properties, for example, nitrogen or boron doping in graphene increases reversible capacity. Ajayan et al.[95] prepared a nitrogen-doped graphene that showed a double reversible discharge capacity. The presence of defects created by nitrogen doping attributes toward the increased reversible discharge capacity. The synergistic effect of heteroatom co-doping can also produce graphene material with high reversible capacity. For example, as reported by Zhang's group, boron and nitrogen co-doped graphene showed a large number of defects that provide active sites for lithium-ion storage, resulting in better reversible capacity for LIBs.[96] In another study, nitrogen and phosphorous co-doped graphene showed a fast sodium ion storage and improved rate performance in SIBs. The enhanced performance by the electrode is explained by high surface area, better conductivity and high capacitive contribution of doped functionalities. In another study, sulfur-doped graphene was used as anode material for SIBs. Doped electrode material showed large number of active sites for Na ion storage and it exhibited stable capacity retention up to 1000 cycles (Fig. 4.4).[97]

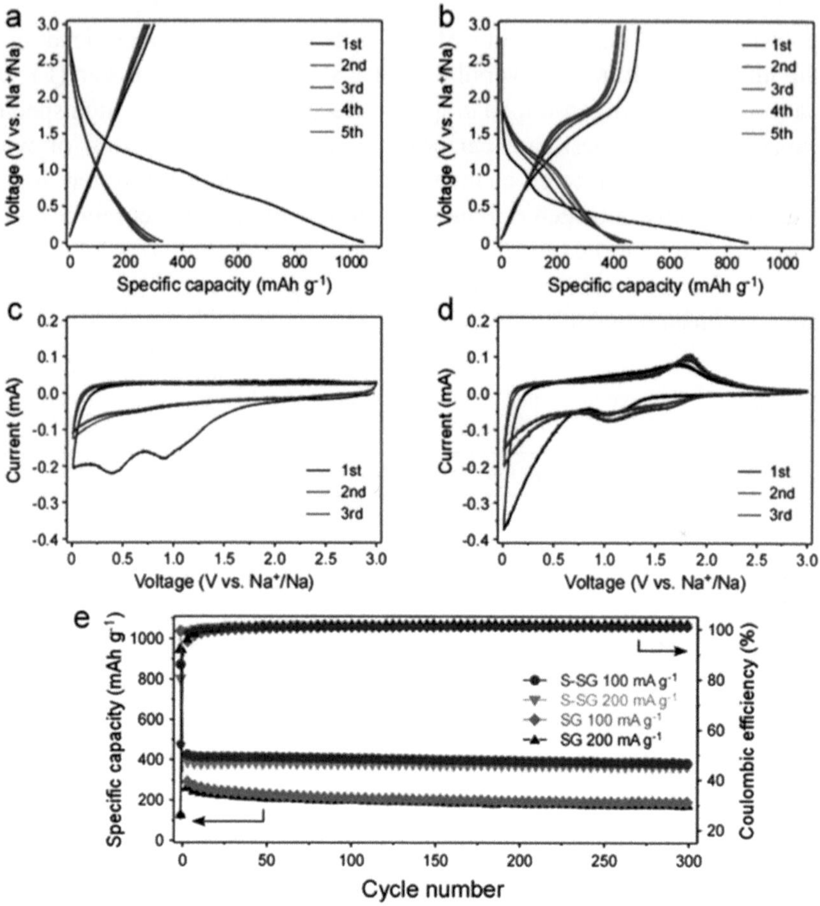

FIGURE 4.4 Voltage profiles of (a) pristine SG and (b) S-SG at a current density of 100 mA g^{-1}. Cyclic voltammograms of (c) pristine SG and (d) S-SG at a scan rate of 0.1 mV s^{-1}. (e) Cycle performance of pristine SG and S-SG at current densities of 100 and 200 mA g^{-1}.[98] (Open access, Permission not required).

In LSBs, nitrogen-doped graphene was employed as a stable, conductive interconnected medium for sulfur cathode. Nitrogen-doped graphene assisted in decreasing polysulfide dissolution into the electrolyte and transport effects during cycling, thereby the electrode showed a better rate capability and cycling stability.[98] Nitrogen and sulfur co-doped graphene demonstrated as an excellent cathode material for lithium–sulfur batteries. The excellent performance was attributed to the synergic effects in which doped graphene can modify the electron distribution and lithium polysulfide binding capability of the N, S functional groups, leading to the improvement in the reversibility of LSBs.[99] Lithium–oxygen (Li–O_2) batteries with nitrogen-doped graphene electrode exhibited a high specific capacity and the best rate performance due to the formation of well-developed interconnected channels and full exposure of nitrogen doping sites. Evidently, nitrogen doping offers a new method for configuring the air electrode materials for high-power and long-life batteries.[100] In Table 4.4, various heteroatom doped/co-doped graphene employed for battery applications are listed.

TABLE 4.4 Heteroatom-Doped/Co-doped Graphene Applied in Various Batteries and their Advantages.

Doped material	Type of battery	Advantages	Ref
Nitrogen doped graphene	Lithium-sulfur	Excellent conductivity, effectively traps the lithium polysulfides with high specific surface area.	101
Boron-doped graphene	Potassium ion	Good electron conductivity and significant charge transfer	102
Sulfur-doped graphene	Sodium ion	long-term cycling stability and ultrahigh reversible capacity	103
Phosphorus-doped graphene	Lithium ion	Highly enhanced cycle and rate capabilities	46
Nitrogen-doped graphene	Lithium-oxygen	Excellent electrochemical performance and increased discharge capacity	104
Nitrogen, boron co-doped graphene	Lithium ion	Exhibits high capacity and good stability	105
Nitrogen, sulfur co-doped graphene	Lithium ion	ultrafast discharge and charge rate capabilities, long-cycling capability	106
Phosphorus and nitrogen co-doped graphene	Lithium ion	Enhanced electrode performance and excellent rate capability	107

Doped material	Type of battery	Advantages	Ref
Fluorine and nitrogen co-doped graphene	Sodium ion	Increases structural defects, provides active sites, and enlarges the graphene interlayer distance, favorable for Na^{+} ion storage capabilities	108

4.8 CONCLUSION

This chapter summarizes investigations on the synthetic approaches in the development of various kinds of heteroatom-doped graphene and their applications in the field of energy storage systems, particularly batteries and supercapacitors. Heteroatom doping can provide graphene with novel chemical, optical, structural, and electronic properties, depending on the percentage of doping, the nature of dopants, and doping combinations. In this regard, by choosing suitable precursors, starting graphene material, reaction time, and temperature, one would be able to make the microstructure and the properties of heteroatom-doped graphene tailored for various applications. Moreover, the possibility of co-doping heteroatom has been recently introduced for the engineering applications. Despite the fact that there is a remarkable development in this area, still, some challenges are associated with targeted control over heteroatom doping in graphene. We believe that heteroatom doping in graphene materials will open up new practical applications in the near future.

KEYWORDS

- **graphene**
- **heteroatom**
- **supercapacitor**
- **battery**
- **energy storage**

REFERENCES

1. Omer, A. M. Energy, Environment and Sustainable Development. *Renew. Sustain. Energy Rev.* **2008,** *12* (9), 2265–2300.
2. Iqbal, S.; Khatoon, H.; Hussain Pandit, A.; Ahmad, S. Recent Development of Carbon Based Materials for Energy Storage Devices. *Mater. Sci. Energy Technol.* **2019,** *2* (3), 417–428.
3. Zhu, Y.; Murali, S.; Cai, W.; Li, X.; Suk, J. W.; Potts, J. R.; Ruoff, R. S. Graphene and Graphene Oxide: Synthesis, Properties, and Applications. *Adv. Mater.* **2010,** *22* (35), 3906–3924.
4. Nonahal, M.; Rastin, H.; Saeb, M. R.; Sari, M. G.; Moghadam, M. H.; Zarrintaj, P.; Ramezanzadeh, B. Epoxy/PAMAM Dendrimer-Modified Graphene Oxide Nanocomposite Coatings: Nonisothermal Cure Kinetics Study. *Prog. Org. Coat.* **2018,** *114*, 233–243.
5. Yarahmadi, E.; Didehban, K.; Sari, M. G.; Saeb, M. R.; Shabanian, M.; Aryanasab, F.; Zarrintaj, P.; Paran, S. M. R.; Mozafari, M.; Rallini, M.; Puglia, D. Development and Curing Potential of Epoxy/Starch-Functionalized Graphene Oxide Nanocomposite Coatings. *Prog. Org. Coat.* **2018,** *119* (December 2017), 194–202.
6. Rao, C. N. R.; Gopalakrishnan, K.; Govindaraj, A. Synthesis, Properties and Applications of Graphene Doped With Boron, Nitrogen and Other Elements. *Nano Today* **2014,** *9* (3), 324–343.
7. Kumar, R.; Sahoo, S.; Joanni, E.; Singh, R. K.; Maegawa, K.; Tan, W. K.; Kawamura, G.; Kar, K. K.; Matsuda, A. Heteroatom Doped Graphene Engineering for Energy Storage and Conversion. *Mater. Today* **2020,** *39*, 47–65.
8. Lv, R.; Terrones, M. Towards New Graphene Materials: Doped Graphene Sheets and Nanoribbons. *Mater. Lett.* **2012,** *78*, 209–218.
9. Jouyandeh, M.; Khadem, S. S. M.; Habibzadeh, S.; Esmaeili, A.; Abida, O.; Vatanpour, V.; Rabiee, N.; Bagherzadeh, M.; Iravani, S.; Saeb, M. R.; Varma, R. S. Quantum Dots for Photocatalysis: Synthesis and Environmental Applications. *Green Chem.* **2021,** *23* (14), 4931–4954.
10. Kaushal, S.; Kaur, M.; Kaur, N.; Kumari, V.; Singh, P. P. Heteroatom-Doped Graphene as Sensing Materials: A Mini Review. *RSC Adv.* **2020,** *10* (48), 28608–28629.
11. Ma, T.; Fan, Q.; Li, X.; Qiu, J.; Wu, T.; Sun, Z. Graphene-based Materials for Electrochemical CO 2 Reduction. *J. CO2 Utili.* **2019,** *30* (May 2018), 168–182.
12. Usachov, D.; Vilkov, O.; Grüneis, A.; Haberer, D.; Fedorov, A.; Adamchuk, V. K.; Preobrajenski, A. B.; Dudin, P.; Barinov, A.; Oehzelt, M.; Laubschat, C.; Vyalikh, D. V. Nitrogen-Doped Graphene: Efficient Growth, Structure, and Electronic Properties. *Nano Lett.* **2011,** *11* (12), 5401–5407.
13. Hou, Z.; Wang, X.; Ikeda, T.; Terakura, K.; Oshima, M.; Kakimoto, M.; Miyata, S. Interplay between Nitrogen Dopants and Native Point Defects in Graphene. *Phys. Rev. B* **2012,** *85* (16), 165439.
14. Wu, P.; Du, P.; Zhang, H.; Cai, C. Microscopic Effects of the Bonding Configuration of Nitrogen-Doped Graphene on its Reactivity Toward Hydrogen Peroxide Reduction Reaction. *Phys. Chem. Chem. Phys.* **2013,** *15* (18), 6920–6928.

15. Pooja R.; Jindal, V. K. Designing Band Gap of Graphene by B and N Dopant Atoms. *RSC Adv.* **2012,** *3* (3), 802–812
16. Wang, H. M.; Wang, H. X.; Chen, Y.; Liu, Y. J.; Zhao, J. X.; Cai, Q. H.; Wang, X. Z. Phosphorus-Doped Graphene and (8, 0) Carbon Nanotube: Structural, Electronic, Magnetic Properties, and Chemical Reactivity. *Appl. Surf. Sci.* **2013,** *273*, 302–309
17. Seredych, M.; Bandosz, T. J. S-Doped Micro/Mesoporous Carbon-Graphene Composites as Efficient Supercapacitors in Alkaline Media. *J. Mater. Chem. A* **2013,** *1* (38), 11717–11727
18. Lin, T.; Huang, F.; Liang, J.; Wang, Y. A Facile Preparation Route for Boron-Doped Graphene, and its CdTe Solar Cell Application. *Energy Environ. Sci.* **2011,** *4* (3), 862–865.
19. Thirumal, V.; Pandurangan, A.; Jayavel, R.; Ilangovan, R. Synthesis and Characterization of Boron Doped Graphene Nanosheets for Supercapacitor Applications. *Synthetic Metals* **2016,** *220*, 524–532.
20. Deng, D.; Pan, X.; Yu, L.; Cui, Y.; Jiang, Y.; Qi, J.; Li, W.-X.; Fu, Q.; Ma, X.; Xue, Q.; Sun, G.; Bao, X. Toward N-Doped Graphene via Solvothermal Synthesis. *Chemi. Mater.* **2011,** *23* (5), 1188–1193.
21. Lou, F.; Buan, M. E. M.; Muthuswamy, N.; Walmsley, J. C.; Ronning, M.; Chen, D. One-Step Electrochemical Synthesis of Tunable Nitrogen-Doped Graphene. *J. Mater. Chem. A* **2015,** *4* (4), 1233–1243.
22. Qu, L.; Liu, Y.; Baek, J. B.; Dai, L. Nitrogen-Doped Graphene as Efficient Metal-Free Electrocatalyst for Oxygen Reduction in Fuel Cells. *ACS Nano* **2010,** *4* (3), 1321–1326.
23. He, D.; Jiang, Y.; Lv, H.; Pan, M.; Mu, S. Nitrogen-Doped Reduced Graphene Oxide Supports for Noble Metal Catalysts with Greatly Enhanced Activity and Stability. *Appl. Catal. B: Environ.* **2013,** *132–133*, 379–388.
24. Lin, Z.; Waller, G. H.; Liu, Y.; Liu, M.; Wong, C. Ping. 3D Nitrogen-Doped Graphene Prepared by Pyrolysis of Graphene Oxide with Polypyrrole for Electrocatalysis of Oxygen Reduction Reaction. *Nano Energy* **2013,** *2* (2), 241–248.
25. Jeong, H. M.; Lee, J. W.; Shin, W. H.; Choi, Y. J.; Shin, H. J.; Kang, J. K.; Choi, J. W. Nitrogen-Doped Graphene for High-Performance Ultracapacitors and the Importance of Nitrogen-Doped Sites at Basal Planes. *Nano Lett.* **2011,** *11* (6), 2472–2477.
26. Jiang, B.; Tian, C.; Wang, L.; Sun, L.; Chen, C.; Nong, X.; Qiao, Y.; Fu, H. Highly Concentrated, Stable Nitrogen-Doped Graphene for Supercapacitors: Simultaneous Doping and Reduction. *Appl. Surf. Sci.* **2012,** *258* (8), 3438–3443.
27. Bai, J.; Zhu, Q.; Lv, Z.; Dong, H.; Yu, J.; Dong, L. Nitrogen-Doped Graphene as Catalysts and Catalyst Supports for Oxygen Reduction in both Acidic and Alkaline Solutions. *Int. J. Hydrog. Energy* **2013,** *38* (3), 1413–1418.
28. Xin, Y.; Liu, J. G.; Jie, X.; Liu, W.; Liu, F.; Yin, Y.; Gu, J.; Zou, Z. Preparation and Electrochemical Characterization of Nitrogen Doped Graphene by Microwave as Supporting Materials for Fuel Cell Catalysts. *Electrochimica Acta* **2012,** *60*, 354–358.
29. Jaison, M. J.; Tharangattu Narayanan, N.; Premkumar, T.; Pillai, V. K. A Single-Step Room-Temperature Electrochemical Synthesis of Nitrogen-Doped Graphene Nanoribbons From Carbon Nanotubes. *J. Mater. Chem. A*, **2015,** *3*, 18222–18228.

30. Li, N.; Wang, Z.; Zhao, K.; Shi, Z.; Gu, Z.; Xu, S. Large Scale Synthesis of N-Doped Multi-Layered Graphene Sheets by Simple Arc-Discharge Method. *Carbon* **2010,** *48* (1), 255–259.
31. Li, X.; Fan, L.; Li, Z.; Wang, K.; Zhong, M.; Wei, J.; Wu, D.; Zhu, H. Boron Doping of Graphene for Graphene-Silicon P-N Junction Solar Cells. *Adv. Energy Mater.* **2012,** *2* (4), 425–429.
32. Sahoo, M.; Sreena, K. P.; Vinayan, B. P.; Ramaprabhu, S. Green Synthesis of Boron Doped Graphene and its Application as High Performance Anode Material in Li Ion Battery. *Mater. Res. Bull.* **2015,** *61*, 383–390.
33. Tennyson, W. D.; Tian, M.; Papandrew, A. B.; Rouleau, C. M.; Puretzky, A. A.; Sneed, B. T.; More, K. L.; Veith, G. M.; Duscher, G.; Zawodzinski, T. A.; Geohegan, D. B. Bottom Up Synthesis of Boron-Doped Graphene for Stable Intermediate Temperature Fuel Cell Electrodes. *Carbon* **2017,** *123*, 605–615.
34. Han, J.; Zhang, L. L.; Lee, S.; Oh, J.; Lee, K. S.; Potts, J. R.; Ji, J.; Zhao, X.; Ruoff, R. S.; Park, S. Generation of B-Doped Graphene Nanoplatelets Using a Solution Process and their Supercapacitor Applications. *ACS Nano* **2013,** *7* (1), 19–26.
35. Panchakarla, L. S.; Subrahmanyam, K. S.; Saha, S. K.; Govindaraj, A.; Krishnamurthy, H. R.; Waghmare, U. V.; Rao, C. N. R. Synthesis, Structure, and Properties of Boron- and Nitrogen-Doped Graphene. *Adv. Mater.* **2009,** *21* (46), 4726–4730.
36. Wu, Z.-S.; Ren, W.; Xu, L.; Li, F.; Cheng, H.-M. Doped Graphene Sheets as Anode Materials with Superhigh Rate and Large Capacity for Lithium Ion Batteries. *ACS Nano* **2011,** *5* (7), 5463–5471.
37. Lü, X.; Wu, J.; Lin, T.; Wan, D.; Huang, F.; Xie, X.; Jiang, M. Low-Temperature Rapid Synthesis of High-Quality Pristine or Boron-Doped Graphene via Wurtz-Type Reductive Coupling Reaction. *J. Mater. Chem.* **2011,** *21* (29), 10685–10689.
38. Gao, H.; Liu, Z.; Song, L.; Guo, W.; Gao, W.; Ci, L.; Rao, A.; Quan, W.; Vajtai, R.; Ajayan, P. M. Synthesis of S-Doped Graphene by Liquid Precursor. *Nanotechnology* **2012,** *23* (27). 275605
39. Jeon, I. Y.; Zhang, S.; Zhang, L.; Choi, H. J.; Seo, J. M.; Xia, Z.; Dai, L.; Baek, J. B. Edge-Selectively Sulfurized Graphene Nanoplatelets as Efficient Metal-Free Electrocatalysts for Oxygen Reduction Reaction: The Electron Spin Effect. *Adv. Mater.* **2013,** *25* (42), 6138–6145.
40. Yang, S.; Zhi, L.; Tang, K.; Feng, X.; Maier, J.; Müllen, K. Efficient Synthesis of Heteroatom (N Or S)-Doped Graphene based on Ultrathin Graphene Oxide-Porous Silica Sheets for Oxygen Reduction Reactions. *Adv. Funct. Mater.* **2012,** *22* (17), 3634–3640.
41. Poh, H. L.; Šimek, P.; Sofer, Z.; Pumera, M. Sulfur-Doped Graphene via Thermal Exfoliation of Graphite Oxide in H2S, SO2, or CS2 Gas. *ACS Nano* **2013,** *7* (6), 5262–5272.
42. Abdelkader-Fernández, V. K.; Domingo-García, M.; López-Garzón, F. J.; Fernandes, D. M.; Freire, C.; López De La Torre, M. D.; Melguizo, M.; Godino-Salido, M. L.; Pérez-Mendoza, M. Expanding Graphene Properties by a Simple S-Doping Methodology Based on Cold CS2 Plasma. *Carbon* **2019,** *144* (2019), 269–279.
43. Li, R.; Wei, Z.; Gou, X.; Xu, W. Phosphorus-Doped Graphene Nanosheets as Efficient Metal-Free Oxygen Reduction Electrocatalysts. *RSC Adv.* **2013,** *3* (25), 9978–9984.

44. Xu, X.; Yang, W.; Chen, B.; Zhou, C.; Ma, X.; Hou, L.; Tang, Y.; Yang, F.; Ning, G.; Zhang, L.; Li, Y. Phosphorus-Doped Porous Graphene Nanosheet as Metal-Free Electrocatalyst for Triiodide Reduction Reaction in Dye-Sensitized Solar Cell. *Appl. Surf. Sci.* **2017,** *405*, 308–315.
45. Chu, K.; Wang, F.; Tian, Y.; Wei, Z. Phosphorus Doped and Defects Engineered Graphene for Improved Electrochemical Sensing: Synergistic Effect of Dopants and Defects. *Electrochimica Acta*, **2017,** *231*, 557–564.
46. Zhang, C.; Mahmood, N.; Yin, H.; Liu, F.; Hou, Y. Synthesis of Phosphorus-Doped Graphene and its Multifunctional Applications for Oxygen Reduction Reaction and Lithium Ion Batteries. *Adv. Mater.* **2013,** *25* (35), 4932–4937.
47. Yao, Z.; Nie, H.; Yang, Z.; Zhou, X.; Liu, Z.; Huang, S. Catalyst-Free Synthesis of Iodine-Doped Graphene via a Facile Thermal Annealing Process and its Use for Electrocatalytic Oxygen Reduction in an Alkaline Medium. *Chem. Commun.* **2012,** *48* (7), 1027–1029.
48. Zhan, Y.; Zhang, B.; Cao, L.; Wu, X.; Lin, Z.; Yu, X.; Zhang, X.; Zeng, D.; Xie, F.; Zhang, W.; Chen, J.; Meng, H. Iodine Doped Graphene as Anode Material for Lithium Ion Battery. *Carbon* **2015,** *94*, 1–8.
49. Chu, K.; Wang, F.; Zhao, X. Lin, Wei, X. Ping, Wang, X. Wei, & Tian, Y. One-Step and Low-Temperature Synthesis of Iodine-Doped Graphene and its Multifunctional Applications for Hydrogen Evolution Reaction and Electrochemical Sensing. *Electrochimica Acta* **2017,** *246*, 1155–1162.
50. Mansour, A. E.; Dey, S.; Amassian, A.; Tanielian, M. H. Bromination of Graphene: A New Route to Making High Performance Transparent Conducting Electrodes with Low Optical Losses. *ACS Appl. Mater. Interfaces* **2015,** *7* (32), 17692–17699.
51. Hasan, M.; Meiou, W.; Yulian, L.; Ullah, S.; Ta, H. Q.; Zhao, L.; Mendes, R. G.; Malik, Z. P.; Ahmad, N. M.; Liu, Z.; Rümmeli, M. H. Direct Chemical Vapor Deposition Synthesis of Large Area Single-Layer Brominated Graphene. *RSC Adv.* **2019,** *9* (24), 13527–13532.
52. Gursu, H.; Gençten, M.; Sahin, Y. Novel Chlorine Doped Graphene Electrodes for Positive Electrodes of a Vanadium Redox Flow Battery. *Int. J. Energy Res.* **2018,** *42* (10), 3303–3314.
53. Wang, K. P.; Zhang, Y. C.; Zhang, X.; Shen, L. Green Preparation of Chlorine-Doped Graphene and its Application in Electrochemical Sensor for Chloramphenicol Detection. *SN Appl. Sci.* **2019,** *1* (2), 1–9.
54. Guo, J.; Zhang, J.; Zhao, H.; Fang, Y.; Ming, K.; Huang, H.; Chen, J.; Wang, X. Fluorine-Doped Graphene with an Outstanding Electrocatalytic Performance for Efficient Oxygen Reduction Reaction in Alkaline Solution. *R. Soc,. Open Sci.* **2018,** *5* (10), 180925.
55. Shen, B.; Chen, J.; Yan, X.; Xue, Q. Synthesis of Fluorine-Doped Multi-Layered Graphene Sheets by Arc-Discharge. *RSC Adv.* **2012,** *2* (17), 6761–6764.
56. Wu, Z. S.; Winter, A.; Chen, L.; Sun, Y.; Turchanin, A.; Feng, X.; Müllen, K. Three-Dimensional Nitrogen and Boron Co-Doped Graphene for High-Performance All-Solid-State Supercapacitors. *Adv. Mater.* **2012,** *24* (37), 5130–5135.
57. Wang, T.; Wang, L. X.; Wu, D. L.; Xia, W.; Jia, D. Z. Interaction between Nitrogen and Sulfur in Co-Doped Graphene and Synergetic Effect in Supercapacitor. *Sci. Rep.* **2015,** *5*, 1–9.

58. Gu, X.; Tong, C. J.; Lai, C.; Qiu, J.; Huang, X.; Yang, W.; Wen, B.; Liu, L. M.; Hou, Y.; Zhang, S. A Porous Nitrogen and Phosphorous Dual Doped Graphene Blocking Layer for High Performance Li-S Batteries. *J. Mater. Chem. A* **2015,** *3* (32), 16670–16678.
59. Yu, X.; Kang, Y.; Park, H. S. Sulfur and Phosphorus Co-Doping of Hierarchically Porous Graphene Aerogels for Enhancing Supercapacitor Performance. *Carbon* **2016,** *101*, 49–56.
60. Huang, S.; Li, Y.; Feng, Y.; An, H.; Long, P.; Qin, C.; Feng, W. Nitrogen and Fluorine Co-Doped Graphene as a High-Performance Anode Material for Lithium-Ion Batteries. *J. Mater. Chem. A* **2015,** *3* (46), 23095–23105.
61. Dong, F.; Cai, Y.; Liu, C.; Liu, J.; Qiao, J. Heteroatom (B, N and P) Doped Porous Graphene Foams for Efficient Oxygen Reduction Reaction Electrocatalysis. *Int. J. Hydrog. Energy* **2018,** *43* (28), 12661–12670.
62. Zhang, J.; Dai, L. Nitrogen, Phosphorus, and Fluorine Tri-Doped Graphene as a Multifunctional Catalyst for Self-Powered Electrochemical Water Splitting. *Angew. Chem.* **2016,** *55* (42), 13296–13300.
63. Thakur, S.; Borah, S. M.; Adhikary, N. C. A DFT Study of Structural, Electronic and Optical Properties of Heteroatom Doped Monolayer Graphene. *Optik* **2018,** *168*, 228–236.
64. Wang, X.; Sun, G.; Routh, P.; Kim, D.-H.; Huang, W.; Chen, P. Heteroatom-Doped Graphene Materials: Syntheses, Properties and Applications. *Chem. Soc. Rev.* **2014,** *43* (20), 7067–7098.
65. Zhu, W.; Gao, J.; Song, H.; Lin, X.; Zhang, S. Nature of the Synergistic Effect of N and S Co-Doped Graphene for the Enhanced Simultaneous Determination of Toxic Pollutants. *ACS Appl. Mater. Interfaces* **2019,** *11* (47), 44545–44555.
66. Velez-Fort, E.; Mathieu, C.; Pallecchi, E.; Pigneur, M.; Silly, M. G.; Belkhou, R.; Marangolo, M.; Shukla, A.; Sirotti, F.; Ouerghi, A. Epitaxial Graphene on 4H-SiC (0001) Grown Under Nitrogen Flux: Evidence of Low Nitrogen Doping and High Charge Transfer. *ACS Nano* **2012,** *6* (12), 10893–10900.
67. Liu, Y.; Feng, Q.; Tang, N.; Wan, X.; Liu, F.; Lv, L.; Du, Y. Increased Magnetization of Reduced Graphene Oxide by Nitrogen-Doping. *Carbon*, **2013,** *60*, 549–551.
68. Khai, T. Van, Na, H. G.; Kwak, D. S.; Kwon, Y. J.; Ham, H.; Shim, K. B.; Kim, H. W. Significant Enhancement of Blue Emission and Electrical Conductivity of N-Doped Graphene. *J. Mater. Chem.* **2012,** *22* (34), 17992–18003.
69. Panchakarla, L. S.; Subrahmanyam, K. S.; Saha, S. K.; Govindaraj, A.; Krishnamurthy, H. R.; Waghmare, U. V.; Rao, C. N. R. Synthesis, Structure, and Properties of Boron- and Nitrogen-Doped Graphene. *Adv. Mater.* **2009,** *21* (46), 4726–4730.
70. Mortazavi, B.; Ahzi, S. Molecular Dynamics Study on the Thermal Conductivity and Mechanical Properties of Boron Doped Graphene. *Solid State Commun.* **2012,** *152* (15), 1503–1507.
71. Denis, P. A. Concentration Dependence of the Band Gaps of Phosphorus and Sulfur Doped Graphene. *Comput. Mater. Sci.* **2013,** *67*, 203–206.
72. Liang, J.; Jiao, Y.; Jaroniec, M.; Qiao, S. Z. Sulfur and Nitrogen Dual-Doped Mesoporous Graphene Electrocatalyst for Oxygen Reduction with Synergistically Enhanced Performance. *Angew. Chem.* **2012,** *51* (46), 11496–11500.

73. Zbořil, R.; Karlický, F.; Bourlinos, A. B.; Steriotis, T. A.; Stubos, A. K.; Georgakilas, V.; Šafářová, K.; Jančík, D.; Trapalis, C.; Otyepka, M. Graphene Fluoride: A Stable Stoichiometric Graphene Derivative and its Chemical Conversion to Graphene. *Small* **2010,** *6* (24), 2885–2891.
74. Zheng, J.; Liu, H.-T.; Wu, B.; Di, C.-A.; Guo, Y.-L.; Wu, T.; Yu, G.; Liu, Y.-Q.; Zhu, D.-B. Production of Graphite Chloride and Bromide using Microwave Sparks. *Sci. Rep.* **2012,** *2* (1), 1–6.
75. Zhang, X.; Hsu, A.; Wang, H.; Song, Y.; Kong, J.; Dresselhaus, M. S.; Palacios, T. Impact of Chlorine Functionalization on High-Mobility Chemical Vapor Deposition Grown Graphene. *ACS Nano* **2013,** *7* (8), 7262–7270.
76. Robinson, J. T.; Burgess, J. S.; Junkermeier, C. E.; Badescu, S. C.; Reinecke, T. L.; Perkins, F. K.; Zalalutdniov, M. K.; Baldwin, J. W.; Culbertson, J. C.; Sheehan, P. E.; Snow, E. S. Properties of Fluorinated Graphene Films. *Nano Lett.* **2010,** *10* (8), 3001–3005.
77. Medeiros, P. V. C.; Mascarenhas, A. J. S.; Mota, F. De B.; de Castilho, C. M. C. A DFT Study of Halogen Atoms Adsorbed on Graphene Layers. *Nanotechnology* **2010,** *21* (48), 485701.
78. Xu, J.; Dong, G.; Jin, C.; Huang, M.; Guan, L. Sulfur and Nitrogen Co-Doped, Few-Layered Graphene Oxide as a Highly Efficient Electrocatalyst for the Oxygen-Reduction Reaction. *Chemsuschem* **2013,** *6* (3), 493–499.
79. Choi, C. H.; Chung, M. W.; Kwon, H. C.; Park, S. H.; Woo, S. I. B, N- and P, N-Doped Graphene as Highly Active Catalysts for Oxygen Reduction Reactions in Acidic Media. *J. Mater. Chem. A* **2013,** *1* (11), 3694–3699.
80. Simon, P.; Gogotsi, Y. Materials for Electrochemical Capacitors. *Nat. Mater.* **2008,** *7* (11), 845–854.
81. Salanne, M.; Rotenberg, B.; Naoi, K.; Kaneko, K.; Taberna, P. L.; Grey, C. P.; Dunn, B.; Simon, P. Efficient Storage Mechanisms for Building Better Supercapacitors. *Nat. Energy* **2016,** *1* (6), 16070.
82. Huang, J.; Sumpter, B. G.; Meunier, V. Theoretical Model for Nanoporous Carbon Supercapacitors. *Angew. Chem.* **2008,** *47* (3), 520–524.
83. Zhou, Y.; Xu, X.; Shan, B.; Wen, Y.; Jiang, T.; Lu, J.; Zhang, S.; Wilkinson, D. P.; Zhang, J.; Huang, Y. Tuning and Understanding the Supercapacitance of Heteroatom-Doped Graphene. *Energy Storage Mater.* **2015,** *1* (September 2018), 103–111.
84. Gopalakrishnan, K.; Govindaraj, A.; Rao, C. N. R. Extraordinary Supercapacitor Performance of Heavily Nitrogenated Graphene Oxide Obtained by Microwave Synthesis. *J. Mater. Chem. A* **2013,** *1* (26), 7563–7565.
85. Niu, L.; Li, Z.; Hong, W.; Sun, J.; Wang, Z.; Ma, L.; Wang, J.; Yang, S. Pyrolytic Synthesis of Boron-Doped Graphene and its Application as Electrode Material for Supercapacitors. *Electrochimica Acta* **2013,** *108*, 666–673.
86. Cheng, L.; Hu, Y.; Qiao, D.; Zhu, Y.; Wang, H.; Jiao, Z. One-Step Radiolytic Synthesis of Heteroatom (N and S) Co-Doped Graphene for Supercapacitors. *Electrochimica Acta* **2018,** *259*, 587–597.
87. Fan, W.; Xia, Y. Y.; Tjiu, W. W.; Pallathadka, P. K.; He, C.; Liu, T. Nitrogen-Doped Graphene Hollow Nanospheres as Novel Electrode Materials for Supercapacitor Applications. *J. Power Sour.* **2013,** *243*, 973–981.

88. Naresh, V.; Bhattacharjee, U.; Martha, S. K. Boron Doped Graphene Nanosheets as Negative Electrode Additive for High-Performance Lead-Acid Batteries and Ultracapacitors. *J. Alloys Comp.* **2019,** *797*, 595–605.
89. Wen, Y.; Wang, B.; Huang, C.; Wang, L.; Hulicova-Jurcakova, D. Synthesis of Phosphorus-Doped Graphene and its Wide Potential Window in Aqueous Supercapacitors. *Chemistry* **2015,** *21* (1), 80–85.
90. Parveen, N.; Ansari, M. O.; Ansari, S. A.; Cho, M. H. Simultaneous Sulfur Doping and Exfoliation of Graphene From Graphite using an Electrochemical Method for Supercapacitor Electrode Materials. *J. Mater. Chem. A* **2015,** *4* (1), 233–240.
91. Jin, T.; Chen, J.; Wang, C.; Qian, Y.; Lu, L. Facile Synthesis of Fluorine-Doped Graphene Aerogel with Rich Semi-Ionic C–F Bonds for High-Performance Supercapacitor Application. *J. Mater. Sci.* **2020,** *55* (26), 12103–12113.
92. Wu, D.; Wang, T.; Wang, L.; Jia, D. Hydrothermal Synthesis of Nitrogen, Sulfur Co-Doped Graphene and its High Performance in Supercapacitor and Oxygen Reduction Reaction. *Microporous and Mesoporous Mater.* **2019,** *290* (June), 109556.
93. Borah, R.; Hughson, F. R.; Johnston, J.; Nann, T. On Battery Materials and Methods. *Mater. Today Adv.* **2020,** *6*, 100046.
94. Nitta, N.; Wu, F.; Lee, J. T.; Yushin, G. Li-Ion Battery Materials: Present and Future. *Mater. Today* **2015,** *18* (5), 252–264.
95. Reddy, A. L. M.; Srivastava, A.; Gowda, S. R.; Gullapalli, H.; Dubey, M.; Ajayan, P. M. Synthesis of Nitrogen-Doped Graphene Films for Lithium Battery Application. *ACS Nano*, **2010,** *4* (11), 6337–6342.
96. Zhang, L.; He, Q.; Huang, S.; Zhu, J.; Key, J.; Shen, P. K. A Novel Boron and Nitrogen Co-Doped Three-Dimensional Porous Graphene Sheet Framework as High Performance Li-Ion Battery Anode Material. *Inorg. Chem. Commun.* **2018,** *96*, 159–164.
97. Quan, B.; Jin, A.; Yu, S.-H.; Kang, S. M.; Jeong, J.; Abruña, H. D.; Jin, L.; Piao, Y.; Sung, Y.-E. Solvothermal-Derived S-Doped Graphene as an Anode Material for Sodium-Ion Batteries. *Adv. Sci.* **2018,** *5* (5), 1700880.
98. Qiu, Y.; Li, W.; Zhao, W.; Li, G.; Hou, Y.; Liu, M.; Zhou, L.; Ye, F.; Li, H.; Wei, Z.; Yang, S.; Duan, W.; Ye, Y.; Guo, J.; Zhang, Y. High-Rate, Ultralong Cycle-Life Lithium/Sulfur Batteries Enabled by Nitrogen-Doped Graphene. *Nano Lett.* **2014,** *14* (8), 4821–4827.
99. Xu, J.; Su, D.; Zhang, W.; Bao, W.; Wang, G. A Nitrogen–Sulfur Co-Doped Porous Graphene Matrix as a Sulfur Immobilizer for High Performance Lithium–Sulfur Batteries. *J. Mater. Chem. A* **2016,** *4* (44), 17381–17393.
100. Zhao, C. T.; Yu, C.; Liu, S. H.; Yang, J.; Fan, X. M.; Huang, H. W.; Qiu, J. 3D Porous N-Doped Graphene Frameworks Made of Interconnected Nanocages for Ultrahigh-Rate and Long-Life Li-O2 Batteries. *Adv. Funct. Mater.* **2015,** *25* (44), 6913–6920.
101. Gao, X. T.; Zhu, X. D.; Gu, L. L.; Wang, C.; Sun, K. N.; Hou, Y. L. Efficient Polysulfides Anchoring for Li-S Batteries: Combined Physical Adsorption and Chemical Conversion In V2O5 Hollow Spheres Wrapped in Nitrogen-Doped Graphene Network. *Chem. Eng. J.* **2019,** *378* (June), 122189.
102. Gong, S.; Wang, Q. Boron-Doped Graphene as a Promising Anode Material for Potassium-Ion Batteries with a Large Capacity, High Rate Performance, and Good Cycling Stability. *J. Phys. Chem. C* **2017,** *121* (44). 24418–24424.

103. Zhao, G.; Yu, D.; Zhang, H.; Sun, F.; Li, J.; Zhu, L.; Sun, L.; Yu, M.; Besenbacher, F.; Sun, Y. Sulphur-Doped Carbon Nanosheets Derived From Biomass as High-Performance Anode Materials for Sodium-Ion Batteries. *Nano Energy* **2020,** *67*, 104219.
104. Li, Y.; Wang, J.; Li, X.; Geng, D.; Banis, M. N.; Li, R.; Sun, X. Nitrogen-Doped Graphene Nanosheets as Cathode Materials with Excellent Electrocatalytic Activity for High Capacity Lithium-Oxygen Batteries. *Electrochem. Commun.* **2012,** *18* (1), 12–15.
105. Huang, S.; Zhang, L.; Zhu, J.; Jiang, S. P.; Shen, P. K. Crumpled Nitrogen- and Boron-Dual-Self-Doped Graphene Sheets as an Extraordinary Active Anode Material for Lithium Ion Batteries. *J. Mater. Chem. A* **2016,** *4* (37), 14155–14162.
106. Wang, Z.-L.; Xu, D.; Wang, H.-G.; Wu, Z.; Zhang, X.-B. In Situ Fabrication of Porous Graphene Electrodes for High-Performance Energy Storage. *ACS Nano,* **2013,** *7* (3), 2422–2430.
107. Ma, X.; Ning, G.; Qi, C.; Xu, C.; Gao, J. Phosphorus and Nitrogen Dual-Doped Few-Layered Porous Graphene: A High-Performance Anode Material for Lithium-Ion Batteries. *ACS Appl. Mater. Interfaces* **2014,** *6* (16), 14415–14422.
108. An, H.; Li, Y.; Gao, Y.; Cao, C.; Han, J.; Feng, Y.; Feng, W. Free-Standing Fluorine and Nitrogen Co-Doped Graphene Paper as a High-Performance Electrode for Flexible Sodium-Ion Batteries. *Carbon,* **2017,** *116*, 338–346.

CHAPTER 5

CARBON NANOTUBES FOR SUSTAINABLE AND CLEAN ENERGY APPLICATIONS

N. G. DIVYA[1], V. N. ANJANA[2], and V. N. ARCHANA[3]

[1]*Cochin University of Science and Technology, Cochin 682022, Kerala, India*

[2]*Sree Sankara Vidyapeetom College, Valayanchirangara, Perumbavoor 683556, Kerala, India*

[3]*Mar Athanasius College, Kothamangalam 686666, Kerala, India*

**Corresponding author. E-mail: archanavn90@gmail.com*

ABSTRACT

This chapter gives an overview of the sustainable and clean energy applications of carbon nanotubes. Carbon nanotubes (CNTs) are elongated tubular structures having diameter of 1–2 nm with the carbon atoms aligned in coaxial cylinders of graphitic sheets. On the basis of the number of graphite layers, carbon nanotubes can be classified into single-walled, double-walled, and multiwalled carbon nanotubes. The electronic structure of CNT is unique so that it provides fascinating applications for a sustainable development. This chapter also deals about the electronic properties and use of CNTs for Li-ion batteries, fuel cells, supercapacitors, and for hydrogen storage, in detail.

Nanostructured Carbon for Energy Generation, Storage, and Conversion. V. I. Kodolov, DSc, Omari Mukbaniani, DSc, Ann Rose Abraham, PhD, and A. K. Haghi, PhD (Eds.)

5.1 CARBON NANOTUBES—TOWARD ENERGY ISSUES—AN INTRODUCTION

Depletion of conventional fossil fuel energy sources and the environmental pollution caused by their consumption has forced scientists to search for eco-friendly renewable energy sources. The increased use of fossil fuels affected our climate badly and also forced people to think about suitable green and sustainable energy sources. Major renewable energy resources such as solar energy, hydro energy, wind energy, and geothermal energy could not fill the gap and thereby there was an urgency for a suitable alternative.[1] Hydrogen, the zero emission fuel is a promising candidate for our sustainable energy applications.[2] Since it is the lightest element exists in gaseous state it is necessary to store it at higher density for the purpose of transportation and various applications. Lack of a safe storage technology makes it difficult for its practical use in our day-to-day life. Nanostructured materials are receiving much scientific attention due to their novel mechanical and electrical behavior. Carbon nanotubes (CNTs) are one such nanostructured material that exhibits these favorable properties. Since the discovery of carbon nanotubes by Iijima in 1991, they are gaining immense focus due to its structural, electronic, chemical, and mechanical properties.[3]

CNTs possess some unique physical properties such as high tensile strength, thermal conductivity, large surface-to-volume ratio, and good electronic properties ranging from semiconductor to metals. The high electron transfer rate over their sidewall leads to higher electron conductivity.[4] The recent and fascinating advancements in materials science have led to the utilization of CNT toward renewable and green energy technologies. The use of carbon-based nanomaterials for viable practical applications is due to their cheap and abundant precursors. Their adsorption properties depend on the surface area and porosity of nanotube. Both sidewall and tube ends of CNTs contribute to the overall electrochemical activity.[5] The electrochemical behavior of the sidewall is similar to that of basal plane of highly oriented pyrolytic graphite, whereas the electrochemical behavior of the tube ends is similar to that of edge plane of highly oriented pyrolytic graphite.[5] Moreover, the adsorption properties can be altered by applying chemical treatments. The intercalation of CNTs into other structures enhanced their performances in efficiency, stability, and durability. Higher cycling efficiency of electrochemical supercapacitors can be achieved by using CNT as conductive filler in the electrode.[5]

What makes CNT special? Along with their confinement providing excellent electrical and mechanical properties, surface properties also contribute to the unique behavior and performance in the field of advanced electronics, energy storage, and conversion. The excellent chemical stability enhances their resistance to the degradation of electrode surface. Owing to the cheap and abundant precursor materials, the use of CNTs for practical applications is economically viable too. The mass production of CNT is a multilevel process, which includes multiple engineering at various levels such as molecules, material, reactor, process, and system-level engineering.[6] Utilization of waste products via recycling is an encouraging step toward the synthesis of CNT. Since the CNT electrode materials can be confined to a small area the electrode–electrolyte contact can be increased. The increase in electrode–electrolyte contact and decrease in the weight of the device maximizes the overall gravimetric performance of the device.

In this article, we review the practical use of CNTs for sustainable and clean energy applications. The superior properties and feasibility possessed by CNTs for the enhanced performances in Li-ion batteries, fuel cells, supercapacitors, and hydrogen storage, which can address the global energy issues that are highlighted.

5.2 ELECTRONIC PROPERTIES OF CNTS

Electronic conduction of a material is an important factor in all energy-related applications. The extraordinary electrical properties of carbon nanotubes can be attributed to the hybridization process of carbon bonds, the unique quasi one-dimensional nature and their cylindrical symmetry. A carbon nanotube consists of one or more coaxial cylindrical graphene sheets rolled up and separated by interlayer of graphite. For single-walled carbon nanotubes (SWCNT), the size distribution is 0.8–3 nm and for multiwalled CNTs (MWCNTs) the size distribution is 2–30 nm in diameter. The electronic properties depend on the structure and the defects present in it. The presence of defects play a profound role in electrical conductivity of CNTs. For single-walled CNT, it must vary in a periodic manner as a function of both helicity and diameter. The nanotube may be metallic or semiconductor depending on the mode of rolling of carbon sheet into a nanotube.[7] The chirality and diameter of a graphite tubule can be denoted by chiral vector

$$C_h = na_1 + ma_2$$

where n and m are integers and a_1 and a_2 are unit vectors of hexagonal honeycomb lattice

The electronic properties of SWCNTs can be explained in terms of their band structure and density of states. This is usually done by considering the electrons in molecular orbitals or by examining using a solid state physics approach, which employs density of states functions and the band structure in the Brillouin zone.[7] Even though the bandgap depends on radius, the presence of defects can alter the situation. Complexity is much more in MWCNTs due to the close proximity of individual nanotubes since the electronic interactions are also taken into account. The orientation of nearby tubes relative to one another. SWCNTs can be classified into three classes depending on their electronic behavior, namely metallic tubules (armchair tubes), semiconducting ones and the third one is semiconducting tubes with moderate bandgaps. The structure can be either metallic or semiconducting depending on the value of integers "n" and "m," though other factors such as bonding, doping, etc. kept the same. Intertubular interactions may occur in bundles of SWCNTs reduces the symmetry of the system and thereby influence the band structure.[7]

The electron mobility of defect-free SWCNT is found to be in the range 7–10 × 10^4 cm^2 V^{-1} s^{-1}, higher than that of nanowires and films. Tailoring the electron transport properties by realizing hybrid CNT-oxide films or nanowires is a recent approach, but not successful till now. This can be attributed to the decrease in electron transfer properties of the hybrid system. High chemical stability, good electrical properties, strong mechanical strength, high aspect ratio, and increased surface area of CNTs makes it suitable for fabricating novel electrodes for clean and sustainable energy. These fascinating properties can enhance the electrode–electrolyte interaction, which promotes ion transport and improve the cycling life of a cell.[8]

5.3 CNTS FOR LITHIUM-ION BATTERIES

Lithium-ion batteries (Li-ion batteries) have been used as a power source in portable electronics for technological breakthroughs to power a wide range of portable devices ranging from portable electronic devices to electronic vehicles that are highly benefited from their high reversible capacity and power capability, good safety, comparatively long cycling life, almost zero-memory effects, environmental-friendliness, and superior energy density over the other well-known rechargeable batteries.[9,10] Li-ion batteries are energy storage devices that rely on the electrochemical cyclic process of energy storage and conversion.[11] The role of electrode materials is vital for the performance of batteries. The recent promising developments in nanotechnology offer great potential prospects to devise novel-nanostructured anode materials for better-performing Li-ion batteries. Nanostructured materials have played a most significant role in the recent progress of Li-ion batteries due to the significant properties of nanostructures such as high surface-area, volume change acclimatization, charge-diffusion span, etc. during the electrochemical charge-discharge process.[9] The nanostructured materials provide favorable electrochemical performance in terms of better energy and power densities that have made Li-ion batteries useful in electronic industries. Extensive research has been performed over the past several decades that has brought substantial scientific progress in battery application.[11]

The energy requirements have soared up with rapid industrialization, forcing the rapid depletion of common nonrenewable resources. This has necessitated mankind, the need for devising new energy sources. Moreover, the unbridled use of nonrenewable resources has imposed a severe threat to the environment. While there has been new-found attention in the areas of renewable and clean sources of energy, and storage of energy. A battery can be used to store the energy, which can be transported and used on requirements.[9] Depending on the reusability, batteries can be generally categorized into two categories such as primary (disposable) and secondary (rechargeable) batteries. Although the primary batteries can be used once as their electrolyte undergoes irreversible changes during discharging, the secondary batteries can be properly discharged and recharged several times. However, the use of portable electronic devices has necessitated the requirements for batteries with comparatively high energy density. Since the commercialization in the 1990s by Sony Corporation, the Li-ion batteries are the leading batteries owing to their high energy density. However, research efforts are in progress to develop ultrathin, flexible, and soft batteries with better charging rate, energy density, operating temperature, power density, safety, durability, and sufficient electrochemical cycling characteristics for their efficient utilisations in portable electronic devices and hybrid vehicles.[9]

Rechargeable type Li-ion batteries are commonly useful for portable electronic devices because of their high voltage, high energy density, lack of memory effect, and long cyclic life. Thus the Li-ion batteries are growing in popularity in aerospace, automotive, and defense applications. The most prominent applications also include electric vehicles and hybrid electric vehicles for transportation, next-generation wireless telecommunication devices, power tools, UPSs, and many more. Despite many of their advantages, further improvement of their properties is essential for the large-scale energy storage applications in hybrid vehicles and smart grids. High cost and safety issues are the two major indispensable drawbacks along with low power density, cyclic performance, and energy density, which further limit their application in these areas.[9]

5.3.1 MECHANISM OF LITHIUM STORAGE

A Li-ion battery comprises three major parts, such as anode, cathode, and conducting electrolyte with good ionic conductivity. In a Li-ion battery, the energy is stored in the electrodes in the form of intercalation compounds of lithium. The positive electrode is known as the cathode that is generally made up of lithium metal oxide and the negative electrode known as the anode is composed of graphitic carbon materials. The electrolyte is usually a lithium salt such as LiBOB (lithium bis(oxalato)borate) or $LiPF_6$ (lithium hexa fluorophosphate) dissolved in ethylene carbonate or dimethyl carbonate. The half-reaction taking place at the cathode and anode can be written as

$$\textbf{\textit{Cathode:}}\ LiMO_2 \leftrightarrow Li_{1-x}MO_2 + xLi^+ + xe^- \tag{5.1}$$

$$\textbf{\textit{Anode:}}\ xLi^+ + xe^- + 6C \leftrightarrow Li_xC_6 \tag{5.2}$$

where, x is the coefficient that is usually one for a complete reaction, and M represents the metal component of lithium metal oxide.[9]

During the charging process, the voltage applied across the electrodes drives the half-reactions in the forward direction from cathode to anode. The metal in the lithium metal oxide is reduced to Li-ions and these ions diffuse through the electrolyte and get into the anode material. The free electrons released during the ion formation are subsequently driven across the external wire connecting the two electrodes to provide the necessary electrons required for the injection of lithium ions into the anode. The voltage needed to conduct this electrochemical process depends upon several factors such as the anode material, type of metal used in lithium metal oxide, and also the type of electrolyte. During the discharging process, the reaction takes place in a reverse direction from anode to cathode and the potential difference generated across the electrodes drives the load (external).[11] This causes the flow of electrons from the anode to the cathode to produce a positive current from the cathode, making the cathode act as the positive terminal of the Li-ion battery.[9]

In a Li-ion battery, the Li-ions are stored by two mechanisms such as intercalation and alloy formation. The intercalation compounds include allotropes of carbon (graphene, CNTs, etc.), TiO_2 compounds, etc., while alloying occurs in metals. The intercalation materials exhibit fast lithium-ion reaction kinetics and structural integrity, which is highly desirable for high-rate capability and long life. While the alloy compounds exhibit high volumetric as well as gravimetric energy densities, and those materials are highly desirable for ultrahigh-power Li-ion batteries. But the major disadvantages include pulverization and electrode failure due to colossal volumetric changes during the lithiation–delithiation process. However, the conductivities of electrode material are vital to facilitate electron conduction during the lithiation–delithiation process. The electrochemical and mechanical behavior made them a promising anode material of Li-ion batteries.[10] The exterior and interior walls of CNTs' are electrochemically active sites for Li intercalation/absorption, justifying the research activities focused on their part as performance boosters of both the electrodes in Li-ion battery.[9]

5.3.2 CONVENTIONAL ANODE MATERIALS

The commercially available Li-ion batteries employ lithium cobalt oxide ($LiCoO_2$) or lithium manganese oxide as cathode material because of its high energy density or higher current density, while the negative electrode (anode) is generally made up of graphitic carbon materials.[9] Graphite has been used commercially since it is comparatively cheap, naturally available, nontoxic, and has a long lifespan.[12] The active materials are mixed with conductive additives and a binder (e.g., polyvinylidene fluoride) in appropriate ratios, before being deposited onto the metal foil. Microporous polymer separators (e.g., polypropylene, polyethylene) are employed to electrically insulate the electrodes, which allow the lithium-ion diffusion.[10] The interstitial spaces in the layered structure of lithiated metal oxides and graphitic carbon atoms made them suitable for the lithium-ion intercalation process. The choice of electrolyte, usually $LiPF_6$ solvated in a mixture of alkyl carbonates (e.g., ethylene carbonate) and dimethyl carbonate (DMC), is also a significant factor for developing a stable electrode–electrolyte interface because of the decomposition of both salt and solvent species of the electrolyte leading to layer components (e.g., LiF, $LiCO_3$).

A major disadvantage of Li-ion batteries at high cycling rates has been the high polarization that reduces the power density, which normally get affected by the diffusion rate of Li-ion in the electrolyte, electrical, and thermal conductivities of the electrode, and the conductivities at the electrode–electrolyte interface.[11] These defects can be overcome by choosing novel electrode materials with high electrical and thermal conductivities, large surface area, or by reducing the ionic diffusion path length. The nanostructured electrodes impart several benefits over the conventional materials such as high reversible Li intercalation capacity without impairing electrode structure, reduced diffusion length that leads to increased lithiation–delithiation rates, enlarged contact area with electrolyte, and reduced volumetric variations during lithiation–delithiation process.[9]

Traditionally, graphite has been the prominent anode material for Li-ion batteries due to its good electrical conductivity arises from the delocalized-bonds, and its proper structure to support Li intercalation and diffusion. Moreover, due to the sp^2 hybridization, the forces between any two adjacent carbon atoms in the same sheet are stronger than those between the adjacent carbon sheets. Thus the Li-ions can get into the interlayer spacing between different graphite sheets, with a little swelling at the site of intercalation, which prevents other Li-ions from occupying adjacent interstitial sites, thus limiting the combination capacity to LiC_6 and specific capacity to 372 mAh g^{-1}.[11] Moreover, the diffusion rate of Li-ion in carbonaceous compounds is between 10^{-12}–10^{-6} cm^2 s^{-1} as against 10^{-9}–10^{-7} for the graphite, resulting in a comparatively low-power battery. Thus, for high-power applications, graphite needs to be replaced with materials having more capacity, energy, and power density. Although the capacity of lithium (3860 mAh g^{-1}) is one of the highest among the different anode materials, the dendrite formation between anode and cathode also raises a security concern. The CNTs offer several advantages over graphite such as high capacity Li-ion batteries and alleviating the risk of pulverization due to their unique morphology, high conductivity, high tensile strength, and low inertness to chemical degradation. Moreover, the SWCNTs are expected to exhibit reversible capacities approximately in the range of 300–600 mAh g^{-1}, effectively superior to that of graphite-based functional properties of CNT-based electrodes.[9]

5.3.3 CNT-BASED COMPOSITE ANODE

Considerable research efforts have been devised recently to examine the most useful electrode materials for Li-ion batteries having comparatively good cycle life, and capacity because the electrode material plays a well-known role in the performance of a battery. As a promising candidate for Li-ion anode materials, the CNTs are one of the most widely investigated carbon-based nanomaterials with good electrochemical potential to be used in Li-ion batteries due to their novel structural, electrical, and mechanical properties.[9] The carbon nanotubes (CNTs) with 1D-tubular structure, large surface area, excellent conductivities, and mechanical flexibility are considered as ideal anode materials to enrich the electrochemical activity chemistry.[9] It has been investigated with regard to Li-ion intercalation process, adsorption, and diffusion of ions by both theoretical and experimental procedures.[11]

The First-principles studies carried out to elucidate the Li-intercalation process in CNTs, show charge transfer between Li and C, and a minor deformation of CNT structures due to intercalation. In addition, both the exterior and interior walls are found as susceptible to Li intercalation. The ab initio calculations aimed at examining Li intercalation through the side-walls or cap regions of CNTs revealed the dependence of barrier for Li insertion on the size of carbon rings. The Li-ion insertion is easier for larger rings and preferred for the interior and exterior of the CNTs. But, the intercalation capacity of CNTs is generally linked to the morphology of the CNTs, which is no longer limited to LiC_6. The defect structure affects the morphologies of the nanotubes and thereby determines their capacities. The defect structure due to the removal of carbon atoms makes "holes" on the walls of carbon nanotubes through which the lithium-ions can easily diffuse into the nanotubes and these ions are free to move within its interior and get accumulated at its exterior. Li-ions can also enter through the ends of the open-ended CNTs. Electron density analyses have demonstrated that complete charge transfer takes place between Li-ions and CNTs as soon as Li intercalates the nanotube. Thus, the major factor dominating the lithium-ion kinetics, in CNTs, is the competency of Li-ions to reach either its open ends or interstitial channel. Moreover, the metallic CNTs are found to have almost 400% more lithium insertion capacity than the semiconducting CNTs due to the conductivity of these CNTs.[9] The combination of CNTs as a conductive additive with the conventional conducting carbons such as graphite or carbon black is an effective strategy to create an electrical percolation network.[10]

5.3.4 MERITS AND DEMERITS AS ANODE OF LI-ION BATTERIES

The CNTs with good electrical conductivities possess promising electrochemical properties making them ideal electrode materials for battery applications. There have been numerous literatures reporting the probable replacement of graphite with CNTs as anode in Li-ions batteries with varying success rate.[9] The CNTs have the potential to get assembled into free-standing electrodes as a promising Li-ion storage material. The CNTs-based anode has greater reversible Li-ion capacities (around 1000 mAh g^{-1}), which is much better than the conventional graphite-based anodes of Li-ion batteries.[10] The use of free-standing CNT-based anodes eliminates the use of copper current collectors, which increases the specific energy density, however, the developmental effort needs to overcome challenges such as charge loss during first cycle, paper crystallinity, etc.[10] It can also provide a good physical support for ultra-high capacity negative electrode materials such as germanium or silicon.[10]

However, the Li-ion insertion through the sidewalls of a CNTs is energetically forbidden, but the open end caps or defective sidewalls usually allow the insertion of lithium-ions inside the CNTs that has been considered as a promising option to enhance the lithiation process.[11] This is because the local energy minima at the interior of CNTs promote its lithium storage efficiency and the potential for rapid diffusion along its length. The open-ended CNTs can be developed by selective oxidation in the presence of an oxidizing acidic medium or ultrasonication procedures.[13] The open-ended SWCNTs exhibited good reversible capacity and favorable voltage profile during lithium extraction that may increase the energy density.[11,13]

Lithium-ion batteries have emerged as a viable candidate for energy requirements because of their better energy density and cyclic capacity in comparison with several other batteries. However, the need for novel nanostructures to improve the performance of conventional electrodes increases with technological advancements. Innovations to overcome most of the challenges associated with the use of CNTs as an efficient electrode material will result in the forthcoming commercialization of the CNTs based batteries

5.4 CNTS FOR HYDROGEN STORAGE

Hydrogen is the most sustainable, safe, and renewable energy carrier, which is expected to replace nonrenewable fossil fuels in near future. The environmental destruction due to the huge consumption of fossil fuels has increased the demand for clean and sustainable energy. In this regard, hydrogen is a promising material to fulfill the energy requirements due to its ability to carry a high gravimetric energy density with almost zero emissions. The main potential of hydrogen lies in powering zero-emission vehicles to reduce atmospheric pollution.[14] Thus, the research effort invested in this area is mainly focused on three things: hydrogen production, storage, and efficient utilization.

Hydrogen has several benefits in comparison with other sources of energy. It is a renewable and clean energy source that generates heat or electricity from chemical reactions, followed by the production of water that can be transformed into hydrogen and oxygen by electrolysis. And the mass storage is easy just as natural gas.[14,15] It is an environmentally friendly source of energy. However, the major challenges have been associated with the production of hydrogen, its storage, and transportation. It can be stored in different ways such as compressed gas, cryogenic liquid, or in solid form, of which the liquid and gaseous types imply the form of hydrogen itself, while the solid-state refers to the form of storage medium.[14,15] In a cryogenic fuel cell, hydrogen is transformed into electricity and water by an electrochemical process without the emission of CO_2 and NO_x. Considering all the drawbacks of the first two storage methods such as cost, various forms of losses of hydrogen, etc., storage of hydrogen in solid form is the most safest and appropriate method to overcome many of these major challenges.

Hydrogen can be stored by absorption on metal hydrides or complex hydrides and adsorption using various carbon-based nanomaterials. But, the adsorption of hydrogen by using carbon-based nanomaterials is more favorable with respect to the storage capacity.[14] The Department of Energy of the USA (US DOE) has proposed a target, according to which the weight storage density must be equal to or greater than 6.5 wt%, and volumetric storage density must be equal to or greater than 63 kg/m^3 up to a pressure of 100 bar.[15] However, the traditional hydrogen storage methods, including high-pressure storage and liquid storage methods cannot meet these requirements at the same time, so several research efforts have been invested to develop more economical and efficient hydrogen storage materials for its practical applications.

5.4.1 BASIC MECHANISM OF HYDROGEN ADSORPTION

The storage of hydrogen in solid form is the safest mode in comparison with the other methods of hydrogen storage. In this method, hydrogen combines with base materials through physisorption or chemisorption processes.[14]

5.4.2 PHYSISORPTION

The physisorption phenomenon of hydrogen adsorption involves the well-known van der Waals interaction force between base materials and hydrogen molecules. The interaction energy is given by the relation, $E = \frac{\alpha H_2 \alpha_{sub}}{R^6}$, where aH2 is the polarizability of H_2, asub is the polarizability of the substrate molecules and *R* is the interaction distance. Thus, to increase the interaction energy, materials with highly polarizable molecules should be chosen as the base. The average energy of interaction between hydrogen molecules and carbon nanomaterials is in the range of 4–5 kJ mol^{-1}.[14]

5.4.3 CHEMISORPTION

Every carbon atom of carbon nanomaterials can act as an interaction site for the chemisorption if the covalent bonding between different carbon atoms of the nanomaterials is effectively used. At high pressure, the hydrogen molecules break up into constituent atoms that result in the development of CH bonds that reduces the distance between two adjacent tubes in CNTs. This may lead to the dissociative adsorption of hydrogen. However, desorption takes place at high temperatures, which is not much useful for practical hydrogen storage applications.[14]

5.4.4 ADSORPTION ENERGY

The process of adsorption and desorption of hydrogen atoms on the surface of carbon nanomaterials depend on the adsorption energy of the nanomaterials. The low adsorption energy describes weak van der Waals force of interactions between the atoms of hydrogen and carbon that may cause desorption of hydrogen at a lower temperature. Thus, the adsorption of hydrogen at ambient temperature may occur only at probably higher pressure. When the adsorption energy is very high, that makes desorption difficult. So, the base materials should be such that they adsorb more hydrogen but the interaction energy can be reduced at the time of desorption. The activation energy also defines the reaction kinetics of the base materials. Though the activation energy is zero during adsorption, a significant barrier needs to be overcome during the desorption process.[14] The hydrogen storage in CNTs is the result of a combined process of physisorption and chemisorption.[15]

5.4.5 CARBON NANOTUBES FOR HYDROGEN STORAGE APPLICATIONS

The development of efficient and economical hydrogen storage systems with higher capacities, lightweight, and good stability is imperative for using hydrogen as a safe and clean future energy carrier. Several disadvantages of metal hydrides, especially huge weight give an upper hand to carbon-based nanomaterials due to their low atomic weight and microporous nature that can adsorb hydrogen molecules at its surface by van der Waal's interactions.[14] A range of carbon nanostructures such as carbon nanotubes (CNTs), activated carbon (AC), carbon nanofibers (CNFs), graphene, graphite, etc. is examined to identify their storage capacity.

The hydrogen storage efficiency of CNTs depends on factors such as structure, geometry, structural defects, temperature, pressure, etc. Both the interior and the exterior of the CNTs, between the CNT tubular structure and shells in the case of MWCNTs, effectively act as the hydrogen storage sites.[14] The hydrogen storage efficiency of CNTs is first reported in 1997 (Dillon et al. in SWCNTs) as 5–10 wt% at 273 Kelvin (0.04 MPa) that could meet the requirements of USA DOE, thereby CNTs are being labeled as an ideal hydrogen storage material. Thereafter, many experimental and theoretical research efforts have been carried out to investigate the hydrogen storage efficiency of SWCNTs and MWCNTs, but the results are inconsistent due to the influence of some internal and external parameters on the storage properties of CNTs.[16–19] The internal parameters include characteristics of CNTs, such as purity, surface area, edge openings, etc., while external factors imply the test gases, measurement methods, temperature, stability of pressure, testing devices, etc.[15]

Nature of the surface, potential energy at the surface, distribution of surface functional groups, surface curvature, and surface area of the adsorbent are important parameters that influence the adsorption process of hydrogen on CNT nanostructures.[18,20–22] The H_2 molecules can be positioned parallel and above the C–C bonds, perpendicular and above the carbon rings, or parallel and above the carbon ring of the CNTs.[18] Though the exact mechanism of hydrogen adsorption is not known, hydrogen can be stored in the interior of CNTs or on their exterior surfaces, or even between the different tubular structures of CNTs especially in MWCNTs. The tube diameter, tube arrangements, spacing between SWCNTs in SWCNT bundles, spacing between wall-to-wall in MWCNTs, wall number of MWCNTs, etc. are also major factors that affect the adsorption efficiency of CNTs.[18]

The temperature and pressure are some important parameters that can affect the adsorption efficiency of CNTs. The research investigation reveals that at a particular temperature, for a given CNT diameter, the adsorption capacity of SWCNTs increases with the rise of operating pressure, whereas it decreases with the rise of operating temperature. But, for a given temperature and pressure, the adsorption capacity of SWCNTs decreases with a decrease in the tubular diameter of CNTs. The bending of CNTs to certain critical angles allows the encapsulation of a large number of hydrogen molecules, and the critical angle decreases with an increase in the length of CNTs. Moreover, the bending of CNTs leads to a change of binding energy between the CNTs and hydrogen by varying the curvature of CNTs, thus significantly affecting the hydrogen adsorption efficiency of CNTs.[23]

The adsorption capacity of CNTs is linearly dependent on the specific surface area of CNTs, pore size, and the defect structure of CNTs. The defect structure not only acts as the entry point of hydrogen but also shortens the diffusion length as well as increases the surface area and pore volume, which in turn will increase the hydrogen adsorption efficiency.[24–26] Moreover, the defect sites create potential wells that enhance the interaction between different CNTs structures and hydrogen, which causes and enhances the adsorption binding energy of H_2. Surface activation, which can be either a physical or chemical process, is an effective method to increase the surface area and porous density. The defect structure and functional groups are also helpful for improving the adsorption efficiency of CNTs. The physical surface treatment method includes ion irradiation, γ irradiation, Compton electron irradiation, mechanical milling, etc. The various chemical treatments process include treatments with acids, bases, and heat. Doping, loading of metal atoms, and heteroatoms are also commonly employed methods to enhance the adsorption efficiency of CNTs.[15] The doping with Pd and Al can effectively improve the adsorption efficiency of CNTs and Co as a dopant improves the cycle stability of the absorption/desorption process. Moreover, B and N can inhibit the metal clustering of metal atoms on CNTs.

5.5 CNTS FOR FUEL CELLS

Chemical energy can be directly converted into electrical energy using a device known as fuel cells. Also these fuel cells enhance its capacity by maintaining low discharges as well as high-energy conversion effectiveness. To improve the efficiency of the fuel cells, Pt catalyst can be used that will stipulate the electrochemical reaction in a fuel cell. Due to the high cost and poor utilization effectiveness of this catalyst, it is better to use a support on the catalyst. These catalytic supported materials provide high surface area for the distribution of the catalyst and also an increase in the electrical conductivity. Some of the catalytic support include carbon (C), and other forms of carbon such as graphene, C nanotubes (CNT), etc.

5.5.1 EFFECT OF CNTS ON FUEL CELL

In addition to other supports CNT supported catalyst exhibits high surface area and electrical conductivity. It is having lightweight, perfect hexagonal structure, and many unfamiliar mechanical, electrical, and chemical properties. PtRu graphitic carbon nanofibers, PtRu CNTs, and PtRu-Vulcan catalysts were used in a silicon micro fuel cell test. The results showed that PtRu nanotubes are having intense power density and PtRu Vulcan with high durability.[27] The use of CNT embraces (i) high catalytic performance, (ii) decreases fuel cell cost, (iii) enhances catalyst steadiness and corrosion resistance, and (iv) increases the transmission capacity.

5.5.1.1 HIGH CATALYTIC PERFORMANCE

Effects of the pretreatment condition using Pt loading as a surface functional groups on multiwalled carbon nanotubes (MWCNTs) provides the polymeric electrolyte membrane fuel cells with high catalytic performance and the amount of oxygen functional groups on the MWCNTs increases as the acid treatment time and temperature increases.[28]

5.5.1.2 DECREASES FUEL CELL COST

The use of carbon nanotubes as a good support provides a large surface area with superior dispersion of platinum thereby it decreases the use of Pt and decrease the whole cost of production.[29]

5.5.1.3 ENHANCES CATALYST STEADINESS AND CORROSION RESISTANCE

The performance and stability of the catalyst as well as corrosion resistance can be improved by the use of CNT. At the same time, the use of CNT decreases the chance of forming the surface oxides and thereby minimizing the corrosion current.[30]

5.5.1.4 INCREASES THE TRANSMISSION CAPACITY

The cathodic quality and electron transmission abilities of Pt catalyst can be enhanced by the use of a mixture of multiwalled carbon nanotubes (MWCNT) and single-walled nanotubes (SWNTs) as Pt supporting materials. When SWNTs are used as a support for a catalyst it will enhance the mass transfer of electrons. In this way, the quality of transmission capacity was analyzed by the implementation of CNT, as a bridge between the catalyst and the current collector.[31]

5.5.2 ENHANCEMENT IN THE PERFORMANCE OF CATALYSTS BY USING CNT

5.5.2.1 CATALYST SUPPORT WITH NITROGEN

For direct methanol fuel cells (DMFCs) several anodic catalysts such as Pt/MnO_2/carbon nanotube (CNT) and PtRu/MnO_2/CNT nanocomposites were developed by incorporating hydrous MnO_2 and Pt and its alloy nanoparticles on CNTs.[32] Electrocatalytic activity and stability for the electro-oxidation of methanol is significantly higher for CNTs functionalized with PtRu nanoparticles supported onto polymeric materials such as poly(diallyldimethylammonium chloride), polyethylenimine, 1-aminopyrene (AP), and tetrahydrofuran.[33]

5.5.2.2 ADDITION OF VARIOUS FORMS OF CARBON

Use of electrode materials in a combined form with graphene and multiwalled CNTs supported fuel cells provides a good catalytic effect.[34] Adding a variety of supporting materials onto CNTs enhances its performance i.e., by doping CNTs with a core-sheath nanostructure, exhibits high durability, high current, and power density.[35] Pt support on multiwalled CNTs doped with nitrogen, act as the cathode catalyst, with an intense power density (0.78 mW/cm^2) than common carbon-supported Pt catalyst (0.72 mW/cm^2).[36] The fuel cell output power for proton exchange is increased to 1.6 times when the CNT is composed with polytetrafluoroethylene, which will reduce the contact resistance between the electrode and plate.[37]

5.5.2.3 *SURFACE TREATMENT ON CNTS*

Surface treatment of CNTs using certain substances, which will be coated by certain nanomaterials as well as polymeric materials enhances its performance. Mesoporous polysulfone substrate on SWNTs leads to the development of 3-D polysulfone—SWCNT, which is a porous anode with high activity for the electron transfer in microbial fuel cells.[38] The CNTs prepared by the adsorption of surfactant on functionalized MWCNTs helps to prevent its reaggregation on nanocomposite and thereby enhances its dispersion and thermal performance in proton exchange.[39]

5.5.2.4 *CATALYST WITH MULTIMETAL ADDITION*

Multiwalled CNTs supported with Pd catalyst can be used as anode for the direct reduction of H_2O_2, and cathode catalysts can be Rh, Ru, Pt, Au, Ag, Pd, Ni, and Cu. The maximum power density of fuel cells increases in the order Cu < Ni < Pd < Rh < Ag < Pt < Ru < Au. Thereby we can explain that multimetal addition on the catalyst has its own effect in the catalytic performance.[40] The multiwalled carbon nanotubes (MWCNT) synthesized via PtRu catalyst with Ni addition, provides a high electrochemical surface of the catalyst.[41]

5.5.2.5 *USE OF ORIENTED CNTS*

When Pt is loaded on oriented CNT it decreases the manufacturing cost and increases the durability. Studies have shown that disordered CNT systems (Pt/CNT) possess less performance than the oriented CNT as cathode toward proton transfer membrane fuel cells.[42] Perpendicularly aligned CNTs with Pt performs good proton, electron, and water transfer. An intense current density and minimizing the use of Pt can be achieved by the use of a catalyst layer having microporous and good conduction in the proton exchange fuel cell.[43] CNTs filled with polyethylene terephthalate and polyvinylidene fluoride form a continuous triple structure that can provide excellent conductivity and strength to bipolar plates in proton exchange membrane fuel cells.[44] The surfactant action can be provided for multiwalled CNTs, by the incorporation of polymeric Nafion film that makes good dispersion and high thermal stability.[45]

5.5.3 INTERACTION BETWEEN CNT AND CATALYST

5.5.3.1 DISPERSION OF CATALYST WITH CNT SUPPORT

For example, PtPd catalyst dispersion can be enhanced by the support of CNT on it. The activity of electrochemical surface area of methanol fuel cells increases when CNT is directly used, which implies that carbon nanostructures can be used as a supporting material to enhance the property. [46,47]

5.5.3.2 DIAMETER AND NUMBER OF LAYERS OF CNT

When oxidation-reduction reaction is carried out using CNT-supported Pt nanoparticles, it is seen that as the diameter of CNT increases the specific surface area and the amount of Pt loading decreases and also the performance of Pt catalyst is more stable.[48] From a comparison in the curve of cathode catalyst using 82 μg Pt/cm^2 in Pt/SWCNT and 12 μg Pt/cm^2 in Pt/MWCNT, it is concluded that multiwalled CNTs provides better performance than single-walled CNTs.[49]

5.5.3.3 ADHESION IN BETWEEN PT NANOPARTICLES AND SUPPORT

The catalytic performance is affected by the shape as well as the size of Pt nanoparticles, which was studied by He et al. Regardless of humidity the adhesive force seems to be stronger as the size of nanoparticles increases. Even though the Pt nanoparticles with tetrahedral geometry are having good adhesion, the shape did not have any significant impact on adhesion property. Superior hydration using Nafion membrane supressed the adhesion and at the same time Nafion membrane helps to improve the absorption.[50]

5.5.4 SYNTHETIC CONDITIONS OF CNTS CATALYST

5.5.4.1 IN SUPERCRITICAL FLUID

CNT-supported Pt catalyst that can be synthesized in supercritical carbon dioxide medium, is a highly effective catalyst for low-temperature fuel cells.[51]

5.5.4.2 USING MICROWAVE TECHNOLOGY

Using microwave-assisted technology functionalized, single-wall carbon nanotubes, can be synthesized in ethylene glycol solution containing H_2PtCl_6, which act as an active catalyst in proton-exchange membrane fuel cells.[52] To synthesize platinum catalyst using the intermittent microwave radiation technology chloroplatinic acid can be used and this catalyst provides better performance than commercial platinum carbon black catalyst.[53]

5.5.4.3 LOW TEMPERATURE PLASMA TECHNOLOGY

Low-temperature plasma technology can be used for the synthesis of CNTs and graphene materials, which can be directly used in fuel cells. The advantage impacts on this technique involves the need of less synthesis time, high activation for the catalyst, less energy needs, high dispersal of active substances, and minimizing the environmental pollution.[54]

5.6 CNTS FOR SUPERCAPACITOR

Supercapacitors can be defined as an electrochemical energy storage device that stores and releases energy through reversible adsorption and desorption of ions at the interfaces that are in between electrode materials and electrolytes. It finds better application for electronic products with high energy density, safety, stability, and low level of heating. Different porous materials, like CNTs, NiO, and Fe–Mn–O composites, can be used to fabricate the electrochemical characteristics.[55] Among these Carbon nanotube-based supercapacitor materials finds excellent performance in its properties Due to their unique properties SWCNTs and MWCNTs have been employed as electrodes in electrochemical supercapacitors.[56,57] Different factors that decide the rate of specific capacitance can be explained.

5.6.1 STRUCTURE EFFECTS

The area of the electrode/electrolyte interface arising by charge carriers depends upon the intensity of electrostatic attraction in EDLC. If the surface area is high and if that area is accessed fully by charge carriers, the higher the capacitance. Higher capacitance does not always depend on surface area. Certain components such as pore size, conductivity, and size distribution also affect the capacitance. A vertically aligned CNTs with 25 nm diameter and 69.5 m^2/g specific area shows an admirable rate capacity with specific capacitance of 14.1 F/g whereas entangled CNT has more conductive paths and large pore structure.[58] NiO/CNT nanocomposites nanoflakes in bundled shape morphology show a maximum specific capacitance of 258 F/g at 1 A/g current density, which is 2.15 times higher than that of bare NiO (120 F/g).[59]

5.6.2 HEATING EFFECTS

Graphitization in CNT can be improved using heating effects. The effect of temperature on growth and the quality as well as the structure of CNTs are the effects that regulate its capacitance. After a heating treatment at 1650°C the capacitance of SWCNTs reduces to 18 F/g from 40 F/g that is due to more perfect graphitization of CNTs. The reason for the decrease in capacitance with increasing temperature is due to the reduction in surface area.[60] At the same time, an advancement on graphitization as a result of heating treatment led to an enhancement in power density and a decrease in equivalent series resistance.[61]

5.6.3 CNT–METAL OXIDE COMPOSITE FOR SUPERCAPACITOR

Full utilization and the advantages of the pseudo capacitance and EDLC can be employed by the incorporation of CNT and oxides of metal. The network formed is open and mesoporous that helps the ions to diffuse easily and as a result there will be a decrease in equivalent series resistance and increase in power density.

5.6.3.1 RUTHENIUM OXIDE–CNTS COMPOSITE

Ruthenium oxide is one among the most promising materials at nanoscale, in energy storage devices. The surface functionality of CNTs affects the electrostatic charge storage and pseudo faradaic reactions of RuO_2–CNT composite.[62] The specific capacitance is shown to be 70 F/g in the case of RuO_2/pristine CNT nanocomposites whereas for RuO_2/hydrophilic CNT nanocomposites it was 120 F/g. A 3D-CNT film with RuO_2 gives a high-rate capability and also a high specific capacitance of 1170 F/g.[63] A key factor to improve the capacitance of metal oxide electrode materials for supercapacitors can be achieved by increasing the dispersion of RuO_2. Also a highly dispersed RuO_2 nanoparticles on nitrogen-containing carbon nanotubes shows an enhanced performance on capacitance.[64]

5.6.3.2 CO_3O_4 AND CNTS COMPOSITE

Co_3O_4, a common metal oxide having great application as anodic material in Li-ion batteries, electrochromic devices, and also as solid-state sensors.[65] An interconnected 3D structure of cobalt oxides nanoparticles with nitrogen-doped carbon nanotubes (N-CNTs) was developed as an effective OER electrocatalyst with large surface area and enhanced charge transfer mechanism. Also, it shows an excellent OER activity and an overpotential of 200 mv, which develops a stable current density of 10 mA cm^{-2}. [66]

5.6.3.3 MANGANESE OXIDE AND CNTS COMPOSITE

One of the most favorable electrode material for specific capacitance with less cost is Manganese oxides. The performance of capacitors using manganese oxide depends on electrolyte PH, which can be based on these two irreversible reactions, Mn(IV)–Mn(II) and Mn(IV)–Mn(VII). An increase in charge–discharge current of 2 A/g with an intense cycle performance is shown by amorphous MnO_2-SWNTs synthesized via room temperature method. Even at a high current of 2 A/g, it shows a best coulombic efficiency of 75% and even after 75 cycles it is having a specific capacitance of 110 F/g.[67] Birnessite-type manganese dioxide uniformly coated on CNT shows an excellent specific capacitance 580 F g^{-1} (320 mAh g^{-1}), which implies that the use of CNTs as a conducting agent enhances the high-rate capability of the MnO_2/CNT composite.[68] MnO_2/MWCNT synthesis via in situ coating method exhibits excellent energy storage capacity with a specific capacitance of 250.5 F/g, which is higher than that of bare MWCNT electrode.[69]

5.6.3.4 $Ni(OH)_2$ AND CNTS COMPOSITE

$Ni(OH)_2$ used in combination with carbon to make hybrid supercapacitors. Here it is known that during the charging process, positive electrode materials converts $Ni(OH)_2$ into NiOOH with the liberation of proton and electron. Thus in $Ni(OH)_2$ system, the diffusion of proton is a measure of its rate capability. $Ni(OH)_2$ /MWCNT composites synthesized via in situ method reduces the aggregation of $Ni(OH)_2$ nanoparticles shows an excellent capacity of 190 mAh g^{-1} at current density of 0.4 A/g. This hybrid material delivered a specific energy of 32 Wh/kg with a power of 1500 W/kg.[70] Carbon nanotube-nickel hydroxide coaxial (CNT/$Ni(OH)_2$) three-dimensional composites provides a high energy density of 35 W h kg^{-1} at a power density of 1.8 kW kg^{-1}, which leads to a favorable path for a high energy and superpower density supercapacitors for the next generation.[71] A polycrystalline $Ni(OH)_2$-CNT composite with a specific capacitance of 544 F g^{-1} shows a high retention capability of 85% even after 1500th cycle with excellent cycling stability.[72]

5.7 CONCLUSION

CNTs are subjected to extensive scientific research to appreciate their potential practical applications since their discovery in 1991. The chirality affects their electronic, thermal, and mechanical properties considerably. The electronic properties of metallic and semiconducting CNTs dramatically differ from each other. However, due to their 1D structure, the electronic transport occurs ballistically, indicating good electrical conductivities. Compared with the properties of other graphitic carbon materials, CNTs possess excellent electrical, mechanical, and thermal properties. Thus CNTs are promising candidates to be explored for use in next generation.

The demands for flexible, lighter, thinner, and higher capacity lithium-ion batteries drive the ongoing research for the development of novel materials with improved properties. Among the choices of different electrode materials, the ability of CNTs to fabricate freestanding electrodes represents a novel approach along with its mechanical and electrochemical properties. Moreover, the opened CNT structure and enriched chirality improve the capacity and electrical transport efficiency of the CNT-based electrodes. Thus, the morphological modification of CNTs provides potential strategies for the improvement of their performance as an anode material

The hydrogen adsorption capacity of CNTs depends on their specific size, geometry, defects, surface area, surface topology, the chemical composition of the surface, micropore volume, adsorption of hydrogen by mesopores, and arrangement of CNTs. The adsorption of hydrogen occurs both inside and outside CNTs, and also at the gap between tubes. And the low operating temperature and high pressure are favorable for the adsorption of hydrogen in CNTs. The doping or surface activation of CNTs with metal oxides or hydroxides, positively charging, mechanical bending of CNTs are different effective strategies to improve the adsorption efficiency of CNTs.

To solve the environmental pollution and future energy crisis fuel cells with high energy conversion efficiency can be used. Carbon nanotubes as catalytic support have a significant role to promises the enhancement in activity of fuel cells. Also CNT-based supercapacitors finds better application due to its superior performance in properties.

In summary, CNTs possess promising uses and potentials in sustainable energy applications. Significant efforts are required to further explore it toward practical applications and commercialization. Further advancements in CNTs will revolutionize our world and provide a clean and sustainable energy in future.

KEYWORDS

- **carbon nanotube**
- **electronic**
- **lithium-ion batteries**
- **fuel cell**
- **supercapacitor**
- **hydrogen storage**
- **sustainable energy**

REFERENCES

1. Liu, C.; Cheng, H.-M. Carbon Nanotubes for Clean Energy Applications. *J. Phys. D: Appl. Phys.* **2005,** *38*, R231.
2. Archana, V.; Rastogi, P. K.; Thoufeeq, S.; Vinayasree, S.; Shaji, S.; Reddy, R.; Garza-Navarro, M. A.; Thomas, S.; Narayanan, T. N.; Anantharaman, M. Engineering Nanostructured Spinel Ferrites by Co-Substitution for Total Water Electrolysis by Preferential Exposure of Metal Cations on the Surface. *Sustain. Energy Fuels* **2020,** *4*, 3915–3925.
3. Schulz, M.; Shanov, V.; Yin, Z.; Cahay, M. *Nanotube Superfiber Materials: Science, Manufacturing, Commercialization*; William Andrew, 2019.
4. Merkoçi, A. Carbon Nanotubes in Analytical Sciences. *Microchimica Acta* **2006,** *152*, 157–174.
5. Tan, C. W.; Tan, K. H.; Ong, Y. T.; Mohamed, A. R.; Zein, S. H. S.; Tan, S. H. Energy and Environmental Applications of Carbon Nanotubes. *Environ. Chem. Lett.* **2012,** *10*, 265–273.
6. Kumar, S.; Nehra, M.; Kedia, D.; Dilbaghi, N.; Tankeshwar, K.; Kim, K.-H. Carbon Nanotubes: A Potential Material for Energy Conversion and Storage. *Prog. Energy Combust. Sci.* **2018,** *64*, 219–253.
7. Centi, G.; Perathoner, S. Carbon Nanotubes for Sustainable Energy Applications. *ChemSusChem* **2011,** *4*, 913–925.
8. Bottani, E. J.; Tascón, J. M. *Adsorption by Carbons*; Elsevier, 2011.
9. Sehrawat, P.; Julien, C.; Islam, S. S. Carbon Nanotubes in Li-Ion Batteries: A Review. *Mater. Sci. Eng. B* **2016,** *213*, 12–40.
10. Landi, B. J.; Ganter, M. J.; Cress, C. D.; DiLeo, R. A.; Raffaelle, R. P. Carbon Nanotubes for Lithium Ion Batteries. *Energy Environ. Sci.* **2009,** *2*, 638–654.
11. Xiong, Z.; Yun, Y. S.; Jin, H. J. Applications of Carbon Nanotubes for Lithium Ion Battery Anodes, *Materials (Basel)* **2013,** *6* (3), 1138–1158.
12. Lim, A. C.; Jadhav, H. S.; Kwon, H. J.; Seo, J. G. High-loading Carbon Nanotubes on Polymer Nanofibers as Stand-Alone Anode Materials for Li-Ion Batteries. *ACS Omega* **2019,** *4* (2), 4129–4137.
13. Shimoda, H.; Gao, B.; Tang, X. P.; Kleinhammes, A.; Fleming, L.; Wu, Y.; Zhou, O. Lithium Intercalation into Opened Single-Wall Carbon Nanotubes: Storage Capacity and Electronic Properties. *Phys. Rev. Lett.* **2002,** *88*, 015502.
14. Mohan, M.; Sharma, V. K.; Kumar, E. A.; Gayathri, V. Hydrogen Storage in Carbon Nanotubes. *Carbon* **2001,** *39* (10), 1447–1454.
15. Lyu, J.; Kudiiarov, V.; Lider, A. An Overview of the Recent Progress in Modifications of Carbon Nanotubes for Hydrogen Adsorption. *Nanomaterials* **2020,** *10*, 255.
16. Mahdizadeh, S. J.; Goharshadi, E. K. Hydrogen Storage on Silicon, Carbon, and Silicon Carbide Nanotubes: A Combined Quantum Mechanics and Grand Canonical Monte Carlo Simulation Study. *Int. J. Hydrog. Energy* **2014,** *39*, 1719–1731.
17. Minami, D.; Ohkubo, T.; Kuroda, Y.; Sakai, K.; Sakai, H.; Abe, M. Structural Optimization of Arranged Carbon Nanotubes for Hydrogen Storage by Grand Canonical Monte Carlo Simulation. *Int. J. Hydrog. Energy* **2010,** *35*, 12398–12404.

18. Gayathri, V.; Devi, N. R.; Geetha, R. Hydrogen Storage in Coiled Carbon Nanotubes. *Int. J. Hydrog. Energy* **2010,** *35*, 1313–1320.
19. Shalabi, A. S.; Halim, W. S. A.; Aal, S. A.; Soliman, K. A. Tuning Hydrogen Storage of Carbon Nanotubes by Mechanical Bending: Theoretical Study. *Mol. Simul.* **2016,** *42*, 709–714.
20. Ioannatos, G. E.; Verykios, X. E. H_2 Storage on Single- and Multi-Walled Carbon Nanotubes. *Int. J. Hydrog. Energy* **2010,** *35*, 622–628.
21. Lee, S. Y.; Park, S. J. Effect of Chemical Treatments on Hydrogen Storage Behaviors of Multi-Walled Carbon Nanotubes. *Mater. Chem. Phys.* **2010,** *124*, 1011–1014.
22. Lee, S. Y.; Rhee, K. Y.; Nahm, S. H.; Park, S. J. Effect of P-Type Multi-Walled Carbon Nanotubes for Improving Hydrogen Storage Behaviors. *J. Solid State Chem.* **2014,** *210*, 256–260.
23. Seenithurai, S.; Pandyan, R. K.; Kumar, S. V.; Saranya, C.; Mahendran, M. Al-Decorated Carbon Nanotube as the Molecular Hydrogen Storage Medium. *Int. J. Hydrog. Energy* **2014,** *39*, 11990–11998.
24. Lee, S. Y.; Park, S. J. Influence of the Pore Size in Multi-Walled Carbon Nanotubes on the Hydrogen Storage Behaviors. *J. Solid State Chem.* **2012,** *194*, 307–312.
25. Rakhi, R. B.; Sethupathi, K.; Ramaprabhu, S. Synthesis and Hydrogen Storage Properties of Carbon Nanotubes. *Int. J. Hydrog. Energy* **2008,** *33*, 381–386.
26. Adeniran, B.; Mokaya, R. Low Temperature Synthesized Carbon Nanotube Superstructures with Superior CO_2 and Hydrogen Storage Capacity. *J. Mater. Chem. A* **2015,** *3*, 5148–5161.
27. Borghei, M.; Scotti, G.; Kanninen, P.; Weckman, T.; Anoshkin, I. V.; Nasibulin, A. G.; Franssila, S.; Kauppinen, E. I.; Kallio, T.; Ruiz, V. Enhanced Performance of a Silicon Microfabricated Direct Methanol Fuel Cell with PtRu Catalysts Supported on Few-Walled Carbon Nanotubes. *Energy* **2014,** *65*, 612–620.
28. Wagner, F.; Lakshmanan, B.; Mathias, M. Electrochemistry and the Future of the Automobile. *J. Phys. Chem. Lett.* **2010,** *1* (14), 2204–2219.
29. Wan, K.; Long, G.-F.; Liu, M.-Y.; Du, L.; Liang, Z.-X.; Tsiakaras, P. Nitrogen-Doped Ordered Mesoporous Carbon: Synthesis and Active Sites for Electrocatalysis of Oxygen Reduction Reaction. *Appl. Catal. B: Environ.* **2015,** *165*, 566–571.
30. Zhao, L., Wang, Z.-B.; Li, J.-L.; Zhang, J.-J.; Sui, X.-L.; Zhang, L.-M. A Newly-Designed Sandwich-Structured Graphene–Pt–Graphene Catalyst with Improved Electrocatalytic Performance for Fuel Cells. *J. Mater. Chem. A* **2015,** *3* (10), 5313–5320.
31. Choi, J.-Y.; Higgins, D.; Chen, Z. Highly Durable Graphene Nanosheet Supported Iron Catalyst for Oxygen Reduction Reaction in PEM Fuel Cells. *J. Electrochem. Soc.* **2011,** *159* (1), B86.
32. Zhou, C.; Wang, H.; Peng, F.; Liang, J.; Yu, H.; Yang, J. MnO_2/CNT Supported Pt and PtRu Nanocatalysts for Direct Methanol Fuel Cells. *Langmuir* **2009,** *25* (13), 7711–7717.
33. Cheng, Y., Xu, C.; Shen, P. K. Effect of Nitrogen-Containing Functionalization on the Electrocatalytic Activity of PtRu Nanoparticles Supported on Carbon Nanotubes for Direct Methanol Fuel Cells. *Appl. Catal. B: Environ.* **2014,** *158*, 140–149.

34. Jha, N.; Jafri, R. I.; Rajalakshmi, N.; Ramaprabhu, S. Graphene-Multi Walled Carbon Nanotube Hybrid Electrocatalyst Support Material for Direct Methanol Fuel Cell. *Int. J. Hydrog. Energy* **2011,** *36* (12), 7284–7290.
35. Sa, Y. J.; Park, C.; Jeong, H. Y.; Park, S.-H.; Lee, Z.; Kim, K. T.; Park, G.-G.; Joo, S. H. Carbon Nanotubes/Heteroatom-Doped Carbon Core–Sheath Nanostructures as Highly Active, Metal-Free Oxygen Reduction Electrocatalysts for Alkaline Fuel Cells. *Angewandte Chemie* **2014,** *126* (16), 4186–4190.
36. Kanninen, P.; Borghei, M.; Sorsa, O.; Pohjalainen, E.; Esko, I. K.; Ruiz, V.; Kallio, T. Highly Efficient Cathode Catalyst Layer based on Nitrogen-Doped Carbon Nanotubes for the Alkaline Direct Methanol Fuel Cell. *Appl. Catal. B: Environ.* **2014,** *156*, 341–349.
37. Show, Y.; Takahashi, K. Stainless Steel Bipolar Plate Coated with Carbon Nanotube (CNT)/Polytetrafluoroethylene (PTFE) Composite Film for Proton Exchange Membrane Fuel Cell (PEMFC). *J. Power Sour.* **2009,** *190* (2), 322–325.
38. Nguyen, T.-H.; Yu, Y.-Y.; Wang, X.; Wang, J.-Y.; Song, H. A 3D Mesoporous Polysulfone–Carbon Nanotube Anode for Enhanced Bioelectricity Output in Microbial Fuel Cells. *Chem. Commun.* **2013,** *49* (91), 10754–10756.
39. Sayeed, M. D.; Kim, Y. H.; Park, Y.; Gopalan, A. I.; Lee, K.-P.; Choi, S.-J. Non-Covalent Bonding Interaction of Surfactants with Functionalized Carbon Nanotubes in Proton Exchange Membranes for Fuel Cell Applications. *J. Nanosci. Nanotechnol.* **2013,** *13* (11), 7424–7429.
40. Oh, T. H.; Jang, B.; Kwon, S. Electrocatalysts Supported on Multiwalled Carbon Nanotubes for Direct Borohydride–Hydrogen Peroxide Fuel Cell. *Int. J. Hydrog. Energy* **2014,** *39* (13), 6977–6986.
41. Valenzuela-Muñiz, A. M.; Alonso-Nuñez, G.; Botte, G. G.; Miki-Yoshida, M.; Verde-Gómez, Y. Influence of Nickel on the Electrochemical Activity of PtRu/ Multiwalled Carbon Nanotubes Electrocatalysts for Direct Methanol Fuel Cells. *J. Appl. Electrochem.* **2014,** *44* (6), 695–700.
42. Li, W.; Wang, X.; Chen, Z.; Waje, M.; Yan, Y. Carbon Nanotube Film by Filtration as Cathode Catalyst Support for Proton-Exchange Membrane Fuel Cell. *Langmuir* **2005,** *21* (21), 9386–9389.
43. Murata, S.; Imanishi, M.; Hasegawa, S.; Namba, R. Vertically Aligned Carbon Nanotube Electrodes for High Current Density Operating Proton Exchange Membrane Fuel Cells. *J. Power Sour.* **2014,** *253*, 104–113.
44. Wu, M.; Shaw, L. L. A Novel Concept of Carbon-Filled Polymer Blends for Applications in PEM Fuel Cell Bipolar Plates. *Int. J. Hydrog. Energy* **2005,** *30* (4), 373–380.
45. Kim, Y. H.; Sayeed, M. D.; Park, Y.; Gopalan, A. I.; Lee, K.-P.; Choi, S.-J. Functionalized Carbon Nanotubes Reinforced Polymer Electrolyte Membranes Prepared by Surfactant-Assisted Casting for Fuel Cell Applications. *J. Nanoelectron. Optoelectron.* **2013,** *8* (6), 551–556.
46. Wang, H.; Bhalchandra, A. K.; Tamaki, T.; Yamaguchi, T. Synthesis of 3D Graphite Oxide-Exfoliated Carbon Nanotube Carbon Composite and its Application as Catalyst Support for Fuel Cells. *J. Power Sour.* **2014,** *260*, 338–348.

47. Girishkumar, G.; Timothy, D. H.; Vinodgopal, K.; Prashant, V. K. Single Wall Carbon Nanotube Supports for Portable Direct Methanol Fuel Cells. *J. Phys. Chem. B* **2006,** *110* (1), 107–114.
48. Al-Youbi, A. O.; Gómez de la Fuente, J. L.; Pérez-Alonso, F. J.; Obaid, A. Y.; Fierro, J. L. G.; Peña, M. A.; Salam, M. A.; Rojas, S. Effects of Multiwalled Carbon Nanotube Morphology on the Synthesis and Electrocatalytic Performance of Pt Supported by Multiwalled Carbon Nanotubes. *Appl. Catal. B: Environ.* **2014,** *150*, 21–29.
49. Carmo, M.; Paganin, V. A.; Rosolen, J. M.; Gonzalez, E. R. Alternative Supports for the Preparation of Catalysts for Low-Temperature Fuel Cells: The use of Carbon Nanotubes. *J. Power Sour.* **2005,** *142* (1–2), 169–176.
50. He, Q.; Joy, D. C.; David, J. K. Nanoparticle Adhesion in Proton Exchange Membrane Fuel Cell Electrodes. *J. Power Sour.* **2013,** *241*, 634–646.
51. Lin, Y.; Cui, X.; Clive, H. Y.; Chien, M. W. PtRu/Carbon Nanotube Nanocomposite Synthesized in Supercritical Fluid: A Novel Electrocatalyst for Direct Methanol Fuel Cells. *Langmuir* **2005,** *21* (24), 11474–11479.
52. Zhang, W.; Chen, J.; Gerhard, F. S.; Ma, Z.-F.; Gordon, G. W. Microwave-Assisted Synthesis of Pt/CNT Nanocomposite Electrocatalysts for PEM Fuel Cells. *Nanoscale* **2010,** *2* (2), 282–286.
53. Soin, N.; Roy, S. S.; Karlsson, L.; McLaughlin, J. A. Sputter Deposition of Highly Dispersed Platinum Nanoparticles on Carbon Nanotube Arrays for Fuel Cell Electrode Material. *Diam. Relat. Mater.* **2010,** *19* (5–6), 595–598.
54. Wang, Q.; Wang, X.; Chai, Z.; Hu, W. Low-Temperature Plasma Synthesis of Carbon Nanotubes and Graphene based Materials and their Fuel Cell Applications. *Chem. Soc. Rev.* **2013,** *42* (23), 8821–8834.
55. Conway, B. E. *Electrochemical Supercapacitors: Scientific Fundamentals and Technological Applications*; Springer Science & Business Media, 2013.
56. Niu, C.; Enid, K. S., Robert Hoch, David Moy, and Howard Tennent. High Power Electrochemical Capacitors based on Carbon Nanotube Electrodes. *Appl. Phys. Lett.* **1997,** *70* (11), 1480–1482.
57. Du, C.; Pan, N. High Power Density Supercapacitor Electrodes of Carbon Nanotube Films by Electrophoretic Deposition. *Nanotechnology* **2006,** *17* (21), 5314.
58. Zhang, H.; Cao, G. P.; Yang, Y. S. Using a Cut–Paste Method to Prepare a Carbon Nanotube Fur Electrode. *Nanotechnology* **2007,** *18* (19), 195607.
59. Sannasi, V.; Maheswari, K. U.; Karthikeyan, C.; Karuppuchamy, S. H_2O_2-Assisted Microwave Synthesis of NiO/CNT Nanocomposite Material for Supercapacitor Applications. *Ionics* **2020,** *26*, 4067–4079.
60. Li, C.; Wang, D.; Liang, T.; Wang, X.; Wu, J.; Hu, X.; Liang, J. Oxidation of Multiwalled Carbon Nanotubes by Air: Benefits for Electric Double Layer Capacitors. *Powder Technol.* **2004,** *142* (2–3), 175–179.
61. An, K. H.; Kim, W. S.; Park, Y. S.; Moon, J.-M.; Bae, D. J.; Lim, S. C.; Lee, Y. S.; Lee, Y. H. Electrochemical Properties of High-Power Supercapacitors using Single-Walled Carbon Nanotube Electrodes. *Adv. Funct. Mater.* **2001,** *11* (5), 387–392.
62. Xie, J.-D.; Patra, J.; Muhammad, A. T. I.; Gandomi, Y. A.; Wu, T.-Y.; Lee, S.-W.; Chang, J.-K. Graphene Induced Crystallinity and Hydrous State Variations of Ruthenium Oxide Electrodes for Superior Energy Storage Performance. *Electrochimica Acta* **2020,** *360*, 136995.

63. Kim, I.-H.; Kim, J.-H.; Kim, K.-B. Electrochemical Characterization of Electrochemically Prepared Ruthenium Oxide/Carbon Nanotube Electrode for Supercapacitor Application. *Electrochem. Solid State Lett.* **2005,** *8* (7), A369.
64. Chung, M.-Y.; Lo, C.-T. High-Performance Binder-Free RuO_2/Electrospun Carbon Fiber for Supercapacitor Electrodes. *Electrochimica Acta* **2020,** *364*, 137324.
65. Guan, H.; Shao, C.; Wen, S.; Chen, B.; Gong, J.; Yang, X. A Novel Method for Preparing Co_3O_4 Nanofibers by using Electrospun PVA/Cobalt Acetate Composite Fibers as Precursor. *Mater. Chem. Phys.* **2003,** *82* (3), 1002–1006.
66. Farid, S.; Qiu, W.; Zhao, J.; Song, X.; Mao, Q.; Ren, S.; Hao, C. Improved OER Performance of Co3O4/N-CNTs Derived from Newly Designed ZIF-67/PPy NTs Composite. *J. Electroanal. Chem.* **2020,** *858*, 113768.
67. Subramanian, V.; Zhu, H.; Wei, B. Synthesis and Electrochemical Characterizations of Amorphous Manganese Oxide and Single Walled Carbon Nanotube Composites as Supercapacitor Electrode Materials. *Electrochem. Commun.* **2006,** *8* (5), 827–832.
68. Ma, S.-B.; Nam, K.-W.; Yoon, W.-S.; Yang, X.-Q.; Ahn, K.-Y.; Oh, K.-H.; Kim, K.-B. Electrochemical Properties of Manganese Oxide Coated onto Carbon Nanotubes for Energy-Storage Applications. *J. Power Sour.* **2008,** *178* (1), 483–489.
69. Xie, X.; Gao, L. Characterization of a Manganese Dioxide/Carbon Nanotube Composite Fabricated using an In Situ Coating Method. *Carbon* **2007,** *45* (12), 2365–2373.
70. Wang, Y.-G.; Yu, L.; Xia, Y.-Y. Electrochemical Capacitance Performance of Hybrid Supercapacitors Based on $Ni(OH)_2$/Carbon Nanotube Composites and Activated Carbon. *J. Electrochem. Soc.* **2006,** *153* (4), A743.
71. Salunkhe, R. R.; Lin, J.; Malgras, V.; Dou, S. X.; Kim, J. H.; Yamauchi, Y. Large-Scale Synthesis of Coaxial Carbon Nanotube/$Ni(OH)_2$ Composites for Asymmetric Supercapacitor Application. *Nano Energy* **2015,** *11*, 211–218.
72. Abitkar, S. B.; Jadhav, P. R.; Tarwal, N. L.; Moholkar, A. V.; Patil, C. E. A Facile Synthesis of α-Ni (OH) 2-CNT Composite Films for Supercapacitor Application. *Adv. Powder Technol.* **2019,** *30* (10), 2285–2292.

CHAPTER 6

CARBON-BASED NANOMATERIALS FOR ENERGY GENERATION AND STORAGE APPLICATIONS

ARUN KUMAR K.V.[1,2*], GREESHMA SARA JOHN[1,2], ATHIRA MARIA JOHNSON[1,2], ARJUN SURESH P.[1,2], and UNNIKRISHNAN N.V.[3]

[1]*Department of Physics, CMS College (Autonomous), Kottayam 686001, Kerala, India*

[2]*Nanotechnology and Advanced Materials Research Centre, CMS College (Autonomous), Kottayam 686001, Kerala, India*

[3]*School of Pure and Applied Physics, Mahatma Gandhi University, Kottayam, Kerala, India*

**Corresponding author. E-mail: arunkumar@cmscollege.ac.in*

ABSTRACT

Carbon-based materials exhibit numerous advantages over other materials due to their properties like high electrical conductivity, low cost, and high surface area. Developing carbon-based materials in the future is an alternating, safe, low-cost, and efficient method in energy storage/generation applications. Carbon can be categorized into different structures like carbon nanotube, fullerene, graphene, etc., and has a wide range of applications. We discussed here the applications of carbon-based nanomaterials devices like supercapacitors, batteries, fuel cells, and photovoltaics in energy storage/generation applications.

Nanostructured Carbon for Energy Generation, Storage, and Conversion. V. I. Kodolov, DSc, Omari Mukbaniani, DSc, Ann Rose Abraham, PhD, and A. K. Haghi, PhD (Eds.)

6.1 INTRODUCTION

Modern society demands massive consumption of energy due to the increased population in the world and fast-growing improved technology needs. However, the excessive use of conventional fossil energy sources creates environmental problems and energy crisis issues. Global energy consumption has been accelerating at an exponential rate. The present consumption rate of global energy exhaustion very soon will become to an end. The effective use of renewable energy devices is necessary in near future with high-performance, low-cost, environment friendly energy conversion, and storage systems. To save our planet and also solve the energy demand, we need more efficient or clean alternative energy sources. In such a situation, carbon-based materials have a great scope in the energy storage/generation sector. Carbon is one of the most abundant materials on earth and is found in different forms like graphite, diamond, and coal. It is one of the highest production materials in technological applications compared to all other elements. Unique hybridization properties, sensitivity to process, and materials properties of nanostructured allotrope forms of carbon have had a great interest in the past two decades. In general, carbon is found in different hybridization states like nanodiamond, fullerene, carbon onion, carbon nanotube (CNT), graphene, etc. The structure and hybridization state of different allotrope shows different chemical, mechanical, thermal, and electrical properties, which opens a wide range of applications in different fields.[1–8] In the 1970s, researchers from Russia and Japan theoretically predicted fullerenes, but in the mid-1980s, experimentally discovered the first fullerene molecule using mass spectroscopy by the laser vaporization of carbon from a graphite target. The name fullerene comes from the famous architect Buckminster Fuller who was an expert in designing and building geodesic domes. In 1991, CNTs were discovered by Ijima which is another allotrope form of carbon with a cylindrical structure. The extraordinary mechanical, electrical, and thermal behavior of CNTs is suitable for a variety of applications. In 2004, another allotrope carbon graphene was discovered by Geim, Novoselov, and co-workers. Graphene consists of a flat monolayer of carbon atoms arranged in a two-dimensional honeycomb structure. The advantage of graphene over CNTs is graphene shows similar or even better mechanical, thermal, and electrical properties. From the engineering point of view, the production and the usage of graphene could be easier when compared to

CNTs.[9–13] Here, we discussed the applications of fullerene, CNTs, and graphene in devices like fuel cells (FCs), photovoltaics (PVs), supercapacitors (SCs), etc.

6.2 GRAPHENE

6.2.1 GRAPHENE SYNTHESIS AND STRUCTURAL PROPERTIES

Graphene is one of the allotropes of carbon and it has a honeycomb structure. Due to its good chemical stability, high surface area, and high conductivity, it is used in various fields of nanoscience. The properties of graphene greatly depend on its preparation methods. We can prepare the material graphene with either by top-down approach or by bottom-up approach. In the top-down approach, graphite is converted into graphene sheets through electrochemical, liquid state oxidation, or by solid-state exfoliations of graphite.[14] In the bottom-up approach, graphene is prepared by building molecular precursors through chemical vapor deposition (CVD), alkaline metal chemical reactions, supersonic spray, etc.[15,41] Generally, the top-down method is considered more economical and suitable for large-scale production. The bottom-up approach is used to prepare high-quality graphene by depositing it on different substrates. The main drawback of the bottom-up approach is its higher cost. CVD method is commonly used in the production of graphene in academic and industrial applications.

Dry exfoliation is one of the simplest methods used to create graphene layers. In this method, layered graphite material is converted into atomic thin sheets by using mechanical or electrostatic force. That is, a few layers of graphene are mechanically peeled from high-oriented pyrolytic graphite.[16,42] The material obtained through this method has high purity and a low density of defects. Even though graphene prepared by this method shows high purity, it is not possible for large-scale production.

Liquid-phase exfoliation (LPE) is another method for synthesizing 2D graphene. It is one of the low-cost techniques for the production of graphene. This method includes three steps:

- Oxidation of graphite.
- Dispersing the oxidized graphite in solvent.
- Reduction of graphite oxide into graphene.

Hummer's method is used to oxidize the graphite by reacting with sodium nitrate in sulfuric acid and potassium permanganate. By stirring, sonification, and thermal expansion, GO flakes can be prepared. But the problem of the LPE method is the use of toxic materials, and this method required a higher boiling point for the production of graphene.

CVD is a method that is used in the large-scale production of graphene sheets. In CVD, the purity of graphene depends on different factors that include temperature, catalyst, gas flow, growth time, etc. Graphene preparation through CVD includes the two steps:

- Pyrolysis of gas.
- Formation of graphene using segregated carbon.

In the first step, the pyrolysis process is used to dissociate carbon atoms, and it must be carried out on the surface of metallic catalysis. All the steps including segregation, decomposition, and pyrolysis require an extreme level of heating of about 2500°C.[3,17] Even though CVD is a very good method for the large-scale production of graphene, the grain boundaries, and ripple formation during synthesis cause defects in graphene.

The facile method discovered in 2008 is another excellent method for fabricating graphene stable aqueous dispersions. This method is also used for making graphene electrodes in batteries and SCs. The aqueous dispersions, thus, obtained can be processed to electrodes using different methods, which include vacuum filtration, dip method, spin coating, inkjet printing, doctor blading, blade coating, etc. Among these methods, vacuum coating is considered the most effective and easy technique. One of the advantages of using this technique is that it does not require any binders and by using this method, we can easily control the device thickness.

6.2.1.1 GRAPHENE

Graphene was first discovered in 2004, and it is one of the allotropes of carbon (Fig. 6.1). It is a flat single sheet of graphite having the ideal two-dimensional (2D) structure with a monolayer of carbon atoms packed into a honeycomb crystal plane.[18] It is the basis of all graphitic materials, such as graphite, fullerene, and CNTs. Today, graphene is recognized to be the thinnest and strongest known material in the universe.[19] It shows excellent electrical, electronic, thermal, mechanical, and optical properties.[20] Both physicists and chemists have speculated about many of its possible potential applications in solar cells, batteries, SCs, flexible displays, sensors, hydrogen storage, and are also used for making many advanced electronic devices such as field-effect transistors, SCs, and sensors. A very important feature of graphene is that the energy of its electrons is linearly dependent on the wave vector near the crossing points within the Brillouin zone.

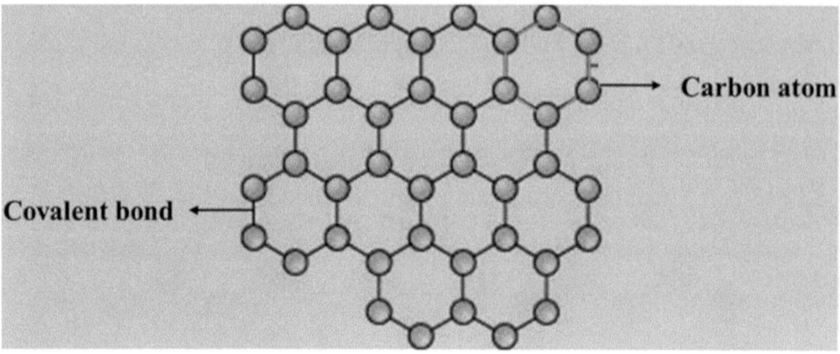

FIGURE 6.1 Structure of graphene.

Graphene exists in three forms—graphene oxide (GO), heteroatom-doped graphene, and 3D graphene. These different forms of graphene exhibit different chemical and physical properties.

6.2.1.2 GRAPHENE OXIDE

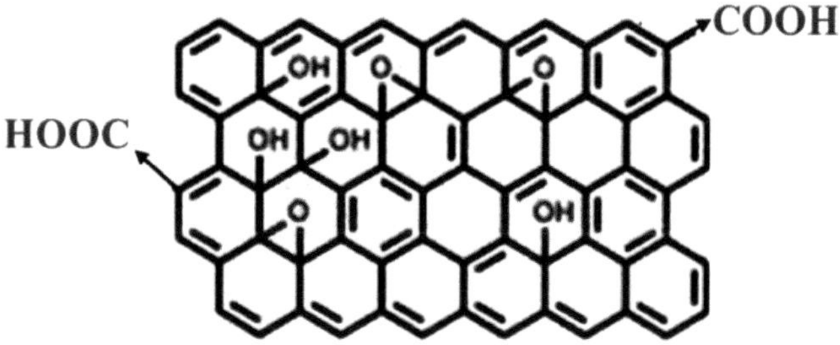

FIGURE 6.2 Structure of GO.

It is a single layer of GO usually produced by the chemical oxidation of graphite (Fig. 6.2). The GO layer shows a proton conductivity of 1.1×10^{-5}–2.8×10^{-3} S cm^{-1} and shows high hydrophilic property, thus it can be applied as electrodes in FCs. Even though it shows good hydrophilic properties, its electrical conductivity is much lower than pure graphene.[21,41] Its application is also limited to low-temperature because of its high sensitivity against temperature in the presence of oxygen.[22,41]

6.2.1.3 HETEROATOM-DOPED GRAPHENE

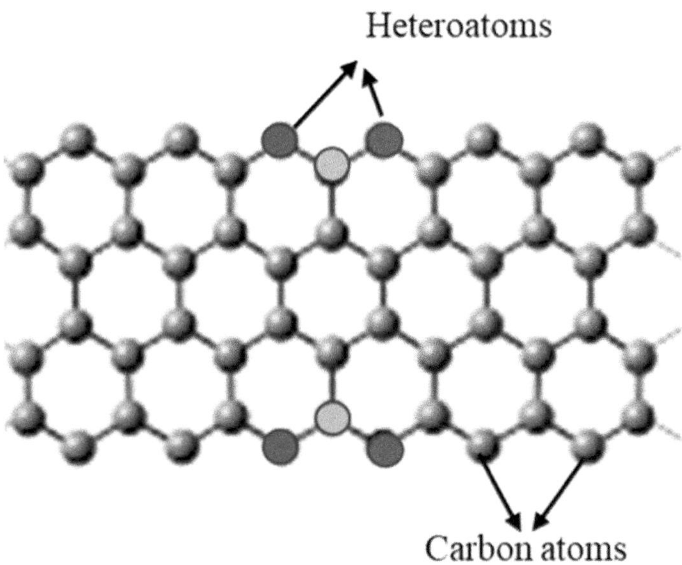

FIGURE 6.3 Structure of heteroatom-doped graphene.

The electronic and chemical properties of graphene can be varied by doping with heteroatoms like B, N, P, S, I, Br, Cl, and F (Fig. 6.3). The donating-withdrawing effects of electron, structural defects can vary its electrical properties. Doping the material with the heteroatoms will also change its density of states near the Fermi energy level and also change its conductivity.[23,41] From the recent studies, it is found that nitrogen-doped graphene shows more stability against chemical oxidation.

6.2.1.4 3D GRAPHENE

One of the problems that exist in using graphene as an electrode in FCs is due to the attractive forces existing between graphene layers. There is a chance of restacking, which will decrease the availability of graphene sites.[24] The issue can be eliminated by using 3D graphene structures. 3D graphene has various architectures that include graphene fiber, graphene sphere, and other nonplanar structures. Changing graphene structure will directly influence the chemical and electrical properties of the material.[25,41]

6.2.2 APPLICATIONS OF GRAPHENE

6.2.2.1 FUEL CELLS

FCs are devices that are similar to batteries but they do not have charging and discharging cycles. In FCs, electricity is generated through an electrochemical reaction. Compared to combustion engines or steam engines which work on fossil fuels, FCs show much more efficiency without hazarding the environment.[2,26] The structure of a typical FC (Fig. 6.4) consists of an electrolyte separated by two electrodes. According to different types of electrolytes, FCs are classified into alkaline FC, phosphoric acid FC, solid-oxide FC, polymer electrolyte membrane fuel cell (PEMFC), molten carbonate FC.

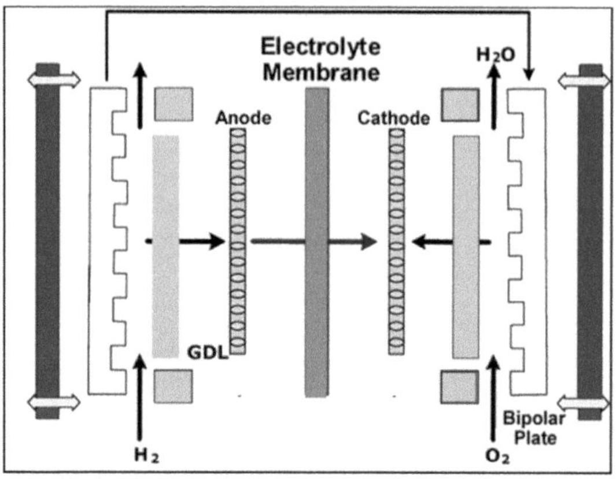

FIGURE 6.4 Fuel cell.

In FCs, platinum electrodes are generally used for cathode and anode, the electrode also acts as a catalyst for speeding up the reaction and can withstand the acidic conditions formed inside the FC. Even though platinum is a good material, but it is very expensive. This is one of the reasons, which prevents its commercialization.

6.2.2.2 GRAPHENE AS ELECTRODES

Graphene is considered as the world's thinnest and strongest material that conducts electricity. Studies found that graphene is a worthy candidate that can be used instead of platinum electrodes. The important properties of graphene include high surface area, high thermal and electrical conductivity, good chemical stability, and high mechanical strength. The theoretical surface area of graphene is about 2630 $m^2\ g^{-1}$, which is much larger than that of graphite.[2,27] These properties of the material graphene make it suitable for FC applications.

6.2.2.3 GRAPHENE AS BIPOLAR PLATES

Bipolar plates in FCs play a significant role in the working of an FC. They are used to conduct heat and current generated, uniformly deliver the air and fuel gas, prevent leakage of gas and coolant in the FC, etc. Generally, nickel, copper, titanium alloys, and steel are used as plates. Even though they show the good conductive property, they undergo corrosion issues. The plates used must have corrosion-resistant properties and should have a high electrical and thermal conductivity at a low cost. Graphene is such a material that can be used in bipolar plates. The corrosion-resistant property of graphene can be applied by coating it with metallic plates. But coating GO or rGO on metallic plates without a binder is challenging, since adding binder will reduce its conductive properties.[2,41]

6.2.2.4 BATTERIES

Energy storage devices are gaining more interest nowadays. Lithium-ion (Li-ion), nickel-cadmium (NiCd), nickel metal hydride (Ni–MH) batteries are mostly used, among this Li-ion batteries show higher energy density per volume and weight. Researches are carried out to increase the energy density of batteries, make batteries transparent, flexible, etc. In the field of energy storage devices also, the material graphene gains more attention due to its high surface area, electrical conductivity, and optical transparent characteristics.[28]

6.2.2.5 GRAPHENE IN LI-ION BATTERIES

Li-ion batteries are one of the foremost popular energy storage devices. Their characteristics which include high energy density, long cycle life, low self-discharging, and low memory effect show their importance.[29] It is made up of four functional components—anode, cathode, electrolyte, and separator. During charging the external power will force the Li-ions to move from cathode to anode through the electrolyte and when the device discharges, the ions will move back to the anode.[31]

6.2.2.6 GRAPHENE AS ELECTRODES

From the recent studies, it is found that graphene and its composite can be used in Li-ion cells (Fig. 6.5) to increase its properties. Using graphene as electrodes that have a higher surface area will increase its storage capacity and its fast-charging capability. This will result in making the batteries more compact and capable. But the issue that is faced by most of the researchers is the difficulty to stabilize the Li-ions with graphene, without defects.[30] Fabricating anode with porous graphene and doped graphene is more suitable for Li-ion batteries.

Using porous graphene instead of planar structured graphene will increase the amount of electrochemically active Li-storage sites.[31] By using graphene as an electrode, the battery becomes lighter and shows fast charging capability.

6.2.2.7 TRANSPARENT BATTERIES

Transparent electronics are one of the main branches that are gaining popularity today. Almost all electronic components including transparent transistors, OLED panels are already in the market. The energy storage device is also an unavoidable source in electronics. Hence its application will further increase if we make it compact and transparent. From the studies, it is found that the electrolyte used in the batteries can be made transparent, but it is difficult in the case of electrodes. Graphene is such a material that can be used as a transparent electrode without decreasing the efficiency of the battery. It is capable of showing a transparency rate of about 97.7%. Good conductivity, large surface area, and transparency make it more suitable for this type of applications.[32]

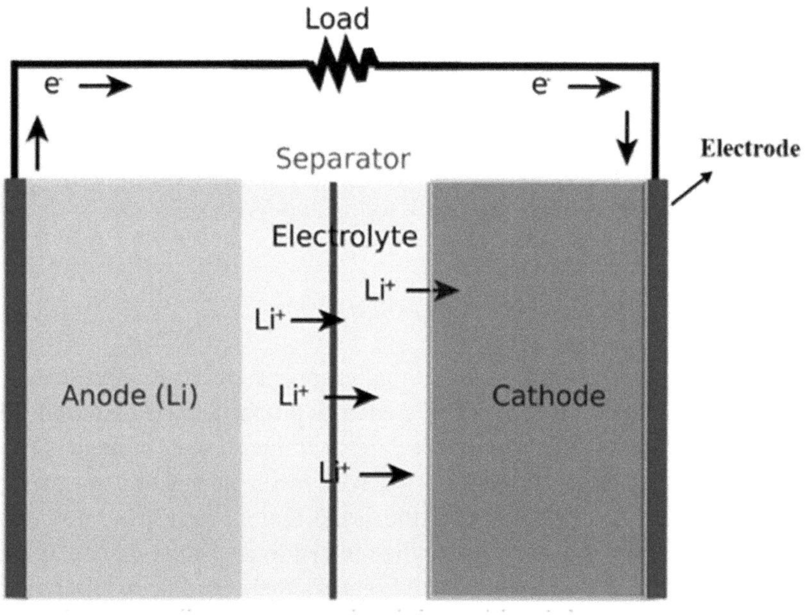

FIGURE 6.5 Li-ion battery.

6.2.2.8 FAST CHARGING BATTERIES

Li-ion batteries are considered one of the best energy storing devices. From the studies, it is found that by using graphene as electrodes in Li-ion batteries, the charging time can be significantly reduced. The graphene electrode can transfer electrons and ions much faster compared to any other electrode. The higher surface area of the graphene electrode also plays a significant role in faster charging and discharging speeds.

6.2.2.9 SUPERCAPACITOR

Energy is generally produced from fossil fuels. For making it more environment friendly, scientists are looking for renewable resources like wind energy and solar energy. However, we face difficulty in storing the energy in a more economic and faster way. SCs are devices that earn more popularity today, compared to conventional batteries, SC has a much faster charging and discharging rate, long life cycle, high power density and shows good performance at low temperature, etc. If SCs can replace conventional batteries, then they can be used for a wide range of applications including energy storage in electric vehicles.

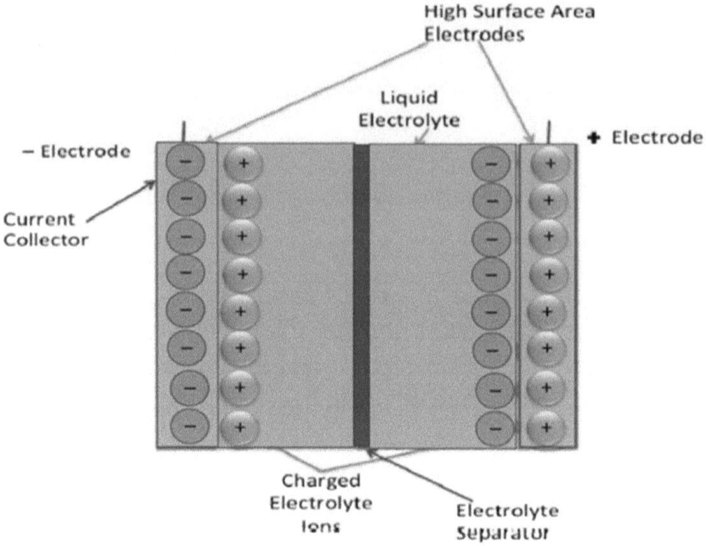

FIGURE 6.6 Supercapacitor.

The main components of SC are electrolyte, electrode, and electrolyte separator (Fig. 6.6). The SC is mainly classified into two types: electric double-layer capacitors (EDLCs) and pseudocapacitors (PCs). EDLC works based on electrostatic interactions, the electrode that is used in the capacitor does not participate in the reaction. The surface area, pore size, and pore structure of the electrode greatly influence its specific capacitance. But in the case of PCs, the electrode used in the capacitor also takes part in the reaction. In this capacitor, energy storage is based on reversible faradaic redox reactions between its electrode and electrolyte.[2,33] Although the conventional capacitors have higher power density, long life cycle, and stability compared to batteries, but its energy density is much lower (5–7 Wh kg^{-1}). Using graphene in an SC can increase the energy density to a much higher value.

6.2.2.10 GRAPHENE AS ELECTRODE IN SUPERCAPACITORS

Various researches are carried out to increase the specific capacitance of SCs, many carbon-based nanomaterials are studied as electrodes including single-walled carbon nanotube (SWCNT). One of the drawbacks of SWCNT is that it has less surface area (1300 m^2 g^{-1}) and high production costs. Graphene is a two-dimensional material that can be used as an electrode in SCs, it has a theoretical surface area of about 2630 m^2 g^{-1}.[34]

6.2.2.11 FLEXIBLE SUPERCAPACITOR USING GRAPHENE AND AG

Currently, capacitors come in various shapes but all are rigid, it will be more useful if the SC becomes flexible. The major problem is that bending of the capacitor causes electrolyte leakage.[28] Graphene is such a good candidate for making flexible SCs that have good electrical properties and a large surface area.

The flexible capacitor that is made from Ag and 3D graphene foam shows a real capacitance of about 38 mF cm^{-2} which is much higher than the reported value.[35]

6.2.2.12 GRAPHENE-BASED STRETCHABLE SUPERCAPACITORS

Stretchable electronics are gaining more popularity in various fields including medicine, robotics, etc. The reason behind this is that it can be deployed in a various environment—on the body, under the skin, etc.[36] Stretchable electronic devices can sustain a large amount of deformation without losing any of their characteristics. Graphene is such a material that can be used to produce stretchable electronics, which includes stretchable SCs. This can be done by attaching graphene paper with an elastomer material.[28] Stretchable capacitors are prepared using the technique of CVD.

6.2.2.13 REPLACING BATTERIES WITH GRAPHENE-BASED SUPERCAPACITORS

Compared to batteries, SCs excel in various fields which include their long-life cycle, fast charging and discharging rates, etc. But the drawback of an SC is its low energy density and if we can increase its energy density then it has the potential to replace conventional batteries. Using graphene in SCs can achieve a similar energy density as that of batteries. Various studies are going on in the field of EDLCs using graphene sheets. But there is a chance of agglomerating the graphene sheets while making electrodes and that will drastically decrease its performance. For solving this issue, a colloidal method is used. In this, a mixture of volatile and nonvolatile are used for the preparation of graphene. The nonvolatile solvent helps to prevent the graphene layers to touch against each other. By using this method, it is also possible to control the density of the electrode.[37]

6.2.2.14 PHOTOVOLTAICS

Graphene is a material that shows different class-leading characteristics that include high surface area, high thermal and electrical conductivity, good chemical stability, and high mechanical strength, high transparency, etc. Due to these properties, graphene can be used for different applications including the field of solar cells (Fig. 6.7).

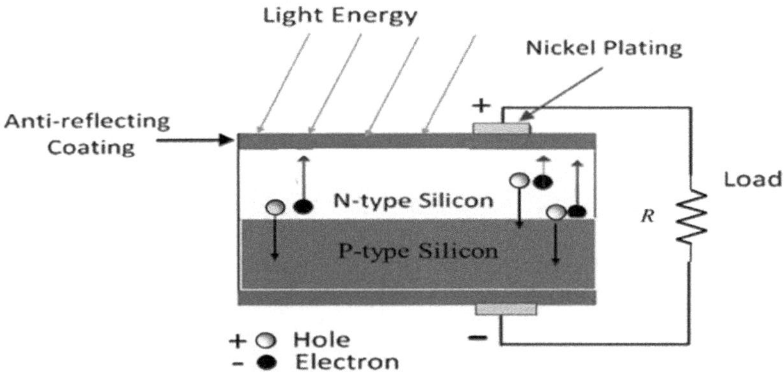

FIGURE 6.7 Solar cell.

6.2.2.15 GRAPHENE-BASED ORGANIC SOLAR CELLS

Generally, in organic solar cells (OSCs) FTO (Fluorine doped tin oxide) or ITO (Indium doped tin oxide) substrates are used as electrodes. The ITO offers transparency of about 80% and offers a resistance of about 10–15 Ω sq^{-1}.[38] The main drawback of ITO substrate is that it is very much sensitive to high and low pH, due to this drawback it reacts with other layers, and leads to low efficiency. The ITO substrates are also very brittle that may form cracks on flexible substrates. Graphene is one of the materials that can solve these issues, its large surface area, high electrical conductivity, and transparency help for this purpose. Graphene layers can be either prepared by using CVD or using Hummer's method. Among these methods, graphene layer prepared by CVD offers better transparency, conductivity, and stability. The transparency of the graphene layer plays an important role in its performance. Transparency of the graphene layer is inversely proportional to its conductance and thickness.

6.2.2.16 USING GRAPHENE AS A HOLE TRANSPORT LAYER (HTL)

The main purpose of HTL in OSC is to conduct holes and block electrons from the active layer. The position of the active layer is in between the donor and acceptor materials so that these layers must be transparent and highly conductive. It is found that direct contact between electrodes and the electron-hole transport layer will cause leakage of current. For avoiding this leakage, an additional functional layer is added in between the active layer and electrode. This layer will help to pass a specific type of charge. In OSC, PEDOT (poly(3,4-ethylenedioxythiophene) is generally used as a hole transporting layer and it offers lesser efficiency due to its resistance. GO is a very good material that can be used as HTL. It offers high efficiency by effectively blocking electrons and by conducting holes.[38]

6.2.2.17 USING GRAPHENE FOR ENCAPSULATION IN PEROVSKITE SOLAR CELLS

Perovskite is considered one of the best materials that can be used for solar cell applications. It exhibits high efficiency which is comparable to silicon solar cells and also its production cost is nearly half that of silicon solar cells. Even though it is cost-effective and it exhibits good properties, there is the main drawback, which is its instability. The stability of perovskite material is very less compared to silicon solar cells. It has been found that silicon solar cell has a lifetime of nearly 25 years but for the perovskite, it is about one year. This is one of the major problems which oppose its commercialization.

Moisture also plays a crucial role in the degradation of performance in perovskite solar cells (PSCs). The moisture in the atmosphere will react with the active layer and affect its stability. By using the encapsulation method, the problem can be solved to some extent. But that will increase the production cost to a great extent. In recent studies, it is found that adding a 2D layer of certain elements will solve the issue. Some materials will exhibit hydrophobic properties against moisture and that will protect the active layer from degradation. Adding a passivation layer between the perovskite and carrier transport layer can improve the device efficiency and wet stability at the same time.[39]

Graphene is one of the substances which shows a hydrophobic effect. The graphene sheet of heterojunction can block the infiltration of O_2 and H_2O molecules and protect the perovskite from degradation in light and humid air. Therefore, assembling a heterojunction from a graphene sheet and perovskite surface is an effective strategy for improving the stability and photoelectric properties of perovskite materials.[40]

6.3 CARBON NANOTUBES (CNT)

6.3.1 CARBON NANOTUBE SYNTHESIS AND STRUCTURAL PROPERTIES

Commonly used procedures for CNT synthesis are as follows:

1. Plasma-based synthesis methods—arc discharge technique, laser ablation technique.
2. Thermal synthesis methods—CVD, plasma-enhanced CVD (PECVD).[43]

6.3.1.1 ARC DISCHARGE TECHNIQUE

It is by arc discharge technique, Iijima first synthesized CNT in 1991.[12,43,47] Also known as the arc evaporation technique, it involves the electrical breakdown of a gas supplied into the chamber to generate plasma when DC is passed through two electrodes (graphite). CNTs are deposited on the cathode. The arc discharge method is best for the synthesis of multiwalled carbon nanotubes (MWCNTs). Arc discharge evaporation of pure graphite rods in hydrogen gas provides an optimum synthesis of MWCNTs with high crystallinity and few synchronized carbon nanoparticles. Using graphite anode-containing metal catalyst yields SWCNTs, but they are obtained from the soot in the gas phase and not from the cathode deposit; maximum yield is achieved with Ni–Co metal catalysts. The growth temperature of this method is higher (above 1700°C) than other CNT synthesis methods which results in the production of nanotubes with a very good degree of crystallinity, which gives better electrical and mechanical properties. The yield per unit time is also higher.[43,47]

6.3.1.2 LASER ABLATION TECHNIQUE

The laser ablation technique is used for the synthesis of high-quality SWCNTs. It was first used by Smalley's group at Rice University in 1995. In this method, vaporization of material (catalyst) from a solid target (graphite) is done by a pulsed or continuous laser in a quartz furnace at 1200°C while an inert gas is blended into the chamber to maintain pressure at 500 Torr. Advantages of this technique include high-quality SWCNT production with lower metallic impurities, diameter control, investigation of growth dynamics, and the production of new materials.[43,47]

6.3.1.3 CHEMICAL VAPOR DEPOSITION (CVD)

CVD is the most common method for the synthesis of CNTs in which a hydrocarbon is thermally decomposed in the presence of a metal catalyst. It is a simple and economic technique that works at low temperature and ambient pressure. This method utilizes a variety of hydrocarbon in any state (solid, liquid, or gas) and enables the use of various substrates. Since it is a low-temperature process (550–1000°C), it offers better control of growth parameters. It enables the growth of CNT in diverse forms such as powder, thin or thick films, aligned or entangled, straight or coiled. Another important advantage is that it even enables the growth of nanotubes of desired architecture at predefined sites on a patterned substrate.[43,47]

6.3.1.4 PLASMA-ENHANCED CVD (PECVD)

PECVD is the most recently investigated synthesis method that works at substantially low temperatures i.e., at room temperatures. A glow discharge generated by a high-frequency voltage is applied to the electrodes in the reaction furnace. This method is mostly known for its ability to produce vertically aligned carbon nanotubes (VACNTs) and is also an appropriate method for the synthesis of CNT hybrid materials and modification of their surface properties.[43]

6.3.1.5 STRUCTURE OF CARBON NANOTUBE

CNT, the elongated fullerene,[47] is a one-dimensional allotropic form of carbon that is most promising and widely studied owing to its exceptional properties. These were first discovered by Sumio Iijima in 1991 as needles have grown at the negative end of the electrode in an arc discharge method for fullerene synthesis. Electron microscopy confirms that each needle comprises coaxial tubes of graphitic sheets i.e. CNT structure is considered to be a graphene sheet rolled into a tube. In CNT, carbon atoms are arranged in a honeycomb lattice, each carbon atom bonded to three neighboring carbon atoms through sp^2 hybridization.[12,43]

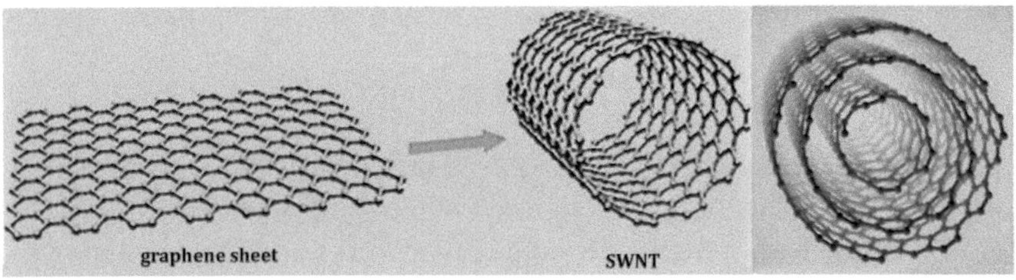

FIGURE 6.8 Single-walled and multiwalled carbon nanotubes.

CNT has a very high aspect ratio (length to diameter ratio) of the order of 100–1000.[45] with length in micrometers and diameter in nanometers. Based upon their lengths CNTs can be long or short CNT (Fig. 6.8). Based on the number of concentric cylindrical layers in the nanostructures CNTs can be of two types—SWCNT and MWCNT. Another type includes open and closed type CNTs where the open type is the cylindrical tube shape with open ends and the closed type is the cylindrical tube shape capped by a half fullerene at the ends. The spiral structure is another type of CNT. There is further another category of CNTs based on crystallographic configurations such as zigzag and armchair and chiral depending on how the sheet of graphene is rolled up along the lattice vectors; the structural parameters being (n,m) diameter and chiral angle. SWCNT is made up of a single atomic layer of graphite forming a cylindrical structure, whereas MWCNT is considered a collection of concentric cylinders of SWCNT with different diameters. Single-walled CNTs may be metallic or semiconducting with a wide bandgap in the range of 0.4–2 eV. Armchair SWCNT exhibits metallic nature, whereas other types of SWCNT exhibit semiconducting nature. MWCNTs are mostly used in nanotube applications. The outer walls of MWCNT protect the inner CNT from chemical interactions which improve the tensile strength which is a nonexistent (partially existent) property in SWCNT. CNTs are also found in the form of nanohorns and nanofibers.[43,44]

CNTs have extraordinary mechanical, thermal, and electrical properties due to their unique hollow structure.[43] One-dimensional morphology, good chemical stability, large surface area, outstanding electrical conductivity, and high current carrying capacity make CNT highly efficient for energy storage applications like SCs.[45,46,48] It also possesses other properties like low density, high ductility, high thermal conductivity, thermal stability, etc., which make them suitable for other applications.[43]

6.3.2 APPLICATIONS OF CARBON NANOTUBE (CNT)

Since the discovery of CNTs in 1991,[12,43,47] they have become quintessential nanomaterials. Their unique structure, outstanding mechanical, thermal and electrical properties increased their importance in many fields of science and technology. CNTs and CNT-based composites have wide area applications including nanoelectronics, PVs, nanosensors, energy storage, and biomedical applications.[60] Higher surface area, good chemical stability, high electrical conductivity, high current carrying capacity, etc., pertain to its application in energy storage devices[46] and PVs.[57] Various energy storage devices include batteries, SCs, FCs,[53] which are renewable alternative solutions for energy storage.

6.3.2.1 BATTERIES AND SUPERCAPACITORS

A lithium-ion battery (LIB) is an electrochemical cell in which the energy transfer occurs across the interface between the electrode and the electrolyte.[49] A SC is also an electrochemical-based energy storage device in which the two electrodes are immersed in an electrolyte solution with a separator in between them and the energy storage is by charge build-up and separation of that charge accumulated onto the two conducting electrodes.[50] The rapid development of microelectronics requires novel LIBs and SCs with high energy densities and large power delivery capabilities.[48] Speaking of LIBs, they are expected to have a high reversible capacity, superior cycling stability, and outstanding rate performance.[49] The intrinsic properties of the electrodes used are an important parameter for the performance of LIBs and SCs. CNTs that have a unique one-dimensional tubular structure, large surface area, high electrical conductivity, good corrosion resistance, high-temperature stability, percolated pore structure, and good mechanical properties can be used as one of the electrodes for enhancing electrochemical properties.[49,50]

CNTs can be used in many ways.

6.3.2.2 CNTS AS ANODES

For high rates of LIBs, electrodes must possess higher electronic/ionic conductivity and higher safety and must be economic. Also, hard carbon material anodes have several disadvantages—extremely high irreversible capacity that consumes a large amount of lithium, energy losses during cycling, and lithium deposition at shallow potentials results in unwanted plating of lithium metals. CNTs with unique structure is the best substitute.[49] The mesoporous structure and large surface area of MWCNTs make them more accessible to ions of electrolyte. Also, MWCNTs offer great chemical and mechanical stability and improved electron transfer kinetics. Thus, CNTs are used as electrodes for SCs.[50,52]

6.3.2.3 CNTS AS LITHIUM STORAGE MATERIAL

The lithium storage of CNTs is determined by structural defects. The existence of morphological defects is important. The amount of lithium that can be stored by CNT is related to the concentration of structural defects it has. Different CNT structures show different capacities. Free-standing CNT-paper and CNT-carbon black mat used as anodes showed high reverse capacity. Also, metal current collectors were not necessary. This pertains to the conductive and self-sustained property of the CNT network. CNT arrays also exhibit high reversible capacity but lower when compared to free-standing CNT-paper and CNT-carbon black mat. CNT arrays show good cycle and rate performance. Mechanical or chemical processes can be used to introduce defects to the CNTs. The ball milling process results in effective shortening in length and opening of the ends of CNT thus lithium diffusion into the CNTs is increased. Spaces within the CNT bundles give extra capacity for lithium storage. Losses during the charging/discharging cycles are a hindrance to efficient lithium storage.[49]

6.3.2.4 CNTS AS CONDUCTIVE ADDITIVES

The addition of CNTs enhances the electrochemical properties of electrodes because many electrode materials are oxides having poor conductivity. The large aspect ratios, high electrical conductivity, excellent mechanical properties, and high chemical stability can be related here. CNTs form an effective continuous and conductive network throughout the composite. A small amount of CNT can improve conductivity and electrochemical properties leading to a higher gravimetric capacity and power density of the electrode. Nonuniform dispersion of CNT can result in heavy polarization losses and severe capacity fading of the composite electrodes. CNT composite electrodes have higher conductivity, higher specific capacity, improved cycling stability, and rate performance than carbon black electrodes.[49]

Liu et al. synthesized MWCNT–$LiMn_2O_4$ nanocomposites as cathodes for LIBS by facile sol–gel method followed by calcination at low temperature. $LiMn_2O_4$ has poor conductivity, severe capacity fading due to the Jahn–Teller distortion at the surface of spinel $LiMn_2O_4$–MWCNT additive provides conducting network that in turn decreases the inner resistance of LIBs, thereby leads to higher specific capacities even at a high charge–discharge current rates. MWCNT–$LiMn_2O_4$ nanocomposites have much higher capacity and smaller polarization during charge–discharge cycles when compared to $LiMn_2O_4$ nanoparticles. Also, the initial discharge capacity and cyclic performance of MWCNT–$LiMn_2O_4$ electrode are higher than those of the $LiMn_2O_4$ electrode. MWCNTs act as intraelectrode wires thereby the charge transfer among the spinel $LiMn_2O_4$ happens with ease. Superior cyclic performance is the result of the excellent electrical conductivity of MWCNTs and the hybridization with the active material.[51]

Chemical inertness of carbon enables in-situ growth of electrode materials on CNTs, which supports the effective fabrication of composite electrodes. Lithium p-block metal alloys can replace graphite in commercial LIBs ($Li_{4.2}Sn$ and $Li_{4.4}Si$ have very high specific capacity).[49]

6.3.2.5 CNTS AS CURRENT COLLECTORS

Current collectors are important components of battery because it holds the electrode material and conducts electricity between the electrode material and the electrode lead. It occupies 15–20% of the total mass of the battery which can reduce the energy density of LIB gravely. They often show less adhesion and limited contact area to electrode materials, which result in reversible capacity fading during the charge and discharge processes. Also, internal impedance may increase due to long-term degradations of metal current collectors. This also causes capacity losses. Properties of CNT like high conductivity, low density, outstanding mechanical properties, high porosity, high surface area, and high chemical stability make it suitable to use as current collectors. Interlaced conductive CNT networks as lightweight current collectors are a better option for next-generation LIBs.[49]

Since the CNT network function as both electrodes and electron transporting channels, conductive additives, and common binder materials are no longer needed. Jiang et al. made use of nickel nanoparticle decorated VACNT array as binder-free electrode for SCs.[49]

Hsia et al. reported highly flexible micro-SC with interdigitated finger electrodes of VACNTs; sputtered Ni to reduce the in-plane resistance of the electrodes and ionogel was used as an electrolyte to synthesize a fully solid-state device.[61]

6.3.2.6 FUEL CELLS

Steady growth in energy demand and exhaustion of fossil fuels make us consider renewable alternative energy devices. The FC is one of many such alternatives. An FC is a device in which a direct conversion of chemical energy to electrical energy takes place. FCs possess high efficacy in energy conversion and low discharges. Platinum (Pt) is the most effective catalyst employed for the electrochemical reactions in FCs. The high price and poor utilization effectiveness of catalyst loading per unit area are its disadvantages. CNTs can be used as supports[54–56] to increase the catalyst's potential. CNTs satisfy the requirements of catalyst support materials—superior specific surface area for the catalyst distribution, higher electrochemical stability under FC functioning conditions, higher electrical conductivity, and chemical stability at suitable temperatures. Also, CNTs are lightweight and have perfect hexagonal formation. Pt–Ru CNTs used in silicon micro-FCs show maximum power density. Pt–Ru alloys catalysts with CNTs as support show efficient dispersion.[54]

Fixing Pt in the inner and outer walls of CNTs enhances the electrocatalytic properties of Pt/CNT structures. Also, CNT catalyst support systems have a higher current density. These properties enhance the efficiency of the FC. The conductive properties of polyaniline polymer used as a conductive polymer in proton-exchange membrane FCs are enhanced by the addition of CNT to polyaniline coating. Thus, CNT improves the electrocatalytic properties of catalysts.[54]

FCs with CNTs as catalyst supports have greater stability and corrosion resistance as the addition of CNTs reduces the formation of surface oxides and corrosion current; thus better durability.[54]

FC production costs can be reduced by decreasing the use of Pt. CNTs support used has a high surface area, excellent dispersion of the Pt, and smaller particle size so that Pt usage can be reduced. Sheng et al. studied the higher catalytic activity of Cu/CuxO with MWCNT supports. They are cheaper substitutes to Pt.[54]

MWCNT enhances anode surface area and volume ratio thus improves the anode electron transmission capability in microbial FCs. The combination of MWCNTs and SWCNTs when used as Pt catalyst supports enhances the cathode quality activities and mass transport in the catalyst layer.[54]

Kim et al. studied Pt-covered MWCNT (Pt NPs/MWCNTs) synthesized via simple synthetic routes as a promising catalyst for proton exchange membrane fuel cells (PEMFCs). This catalyst enhances the oxygen reduction reaction (ORR) and thereby increases the mass activity (ORR kinetic current normalized to Pt weight) of the PEMFC. Acid treatment and annealing of the catalyst further enhance ORR.[56]

6.3.2.7 PHOTOVOLTAICS

One of the most dependable and promising alternative energy sources is solar energy due to its abundance and sustainability. Also, they are environmentally friendly. PV devices do the direct conversion of solar energy to electrical energy. Silicon-based p–n junction solar cells, thin-film solar cells, dye-sensitized solar cells, etc., belong to various generations of solar cells. Silicon-based p–n junction solar cells belong to first-generation solar cells which possess high efficiency and long-term stability. But complicated fabrication and high cost are their drawbacks. CNT films can be incorporated into the traditional silicon solar cells for the generation of silicon-based hybrid heterojunction devices. In contrast to the older silicon p–n junction solar cells, this hybrid device has advantages like simple structure, easy fabrication, promising cell performance, and low cost. CNT/Si solar cells do not require high temperature to achieve diffusion which is not possible for traditional p–n junctions. CNTs are used because of their higher carrier mobility, tunable band-gap and conductivity, desirable optical properties, and excellent flexibility. In this hybrid device, CNT films work as window electrodes to collect holes. The structure of the CNT films and properties such as sheet resistance, purity, and transparency are attributing to the performance of the CNT/Si solar cells. Wei et al. (2007) first reported the CNT/Si solar cells with a power conversion efficiency (PCE) of 1.31%. significant progress over many years resulted in an improved PCE of 15–17%. The CNT films used have a mixture of semiconducting and metallic nanotubes of different diameters and chiralities. From the study done by Tune et al., it was found that metallic SWCNT(m-SWCNT)/Si solar cells showed higher PCE than semiconducting SWCNT(s-SWCNT)/Si solar cells or unsorted SWCNT(u-SWCNT)/Si solar cells. This high PCE pertains to the high optical transparency of the CNT films which make silicon absorb most of the incident light. Introducing an insulating layer of optimum thickness like thin silicon oxide layers increases open-circuit voltage and decreases the saturation current densities that in turn improves the cell performance. A bulk junction (tube–tube junction) was introduced into SWCNT/Si solar cells by Nicola et al.; in that cell, photo-generation takes place at both the silicon layer and SWCNT layer. Using aligned CNT films in solar cells produces a uniform and dense junctions and also provides efficient conducting paths for photo-generated carriers. This result is due to the high aspect ratio, good uniformity of aligned CNT films. CVD synthesized CNT films are composed of MWCNTs or double-walled CNTs which can absorb more light than SWCNTs. CNT films prepared by dry processes have good transparency and conductivity. Changing the film thickness can change the transmittance and sheet resistance. Mixing conducting polymers like PEDOT: PSS with CNT films intensifies junction density, uniformity, and conductivity of CNT/Si solar cells. Inorganic materials like MoO_x can be explored as a hole transport layer because of their good stability and better hole mobility. Studies have shown an increment of 50% in the PCE of MoO_x incorporated solar devices.[57]

Dye-sensitized solar cells (DSSCs) are third-generation solar cells in which dye acts as a photosensitizer. But its efficiency is a big problem. TiO_2 electrode is an important component of DSSC which does the work of electron transport mechanism. Since CNTs have excellent electron-transport properties, modifying TiO_2 with CNTs can result in better efficacy of the DSSCs. A work done by Chang et al. reports that increasing the CNT amount up to an optimum value increases the photocurrent but a further increase of CNT content reduces the photocurrent. Thus, CNTs perform as electron reservoirs to trap electrons and decrease the recombination of electron-hole pairs. Indium tin oxide (ITO) is an important component for DSSC fabrication. ITO is a transparent conductive film (TCFs) on which TiO_2, dye, electrolyte, etc., are coated for DSSC fabrication. CNT film-based transparent electrode can be used as an alternative to ITO due to the advantageous properties of CNT like single layer tubular wall, small diameter, desirable optical and electrical properties, and excellent stability.[58]

Maarouf et al. worked on a graphene–CNT hybrid material to use as an excellent substitute for ITOs. Graphene being a 2D material that is highly transparent, flexible, reasonably chemically stable, and has high mobility combined with CNT is a bonus. Unlike ITOs, graphene–CNT-based TCFs have advantages like low cost, flexibility, high transparency, low sheet resistance, and good electrical and optical properties.[59]

6.4 FULLERENES

6.4.1 *FULLERENE SYNTHESIS AND STRUCTURAL PROPERTIES*

Fullerene, the third allotrope of carbon, was discovered back in 1985 by Harold. W. Kroto, Robert F. Curl, and Richard E. Smalley. Fullerenes denoted as C_{60} consist of 60 carbon atoms which are arranged in 12 pentagons and 20 hexagons.[62] This allotrope owns a structure of truncated icosahedrons and it is named Buckminsterfullerene. The fullerenes are known as the most symmetric molecule since it has 120 symmetry operations done onto itself.[62]

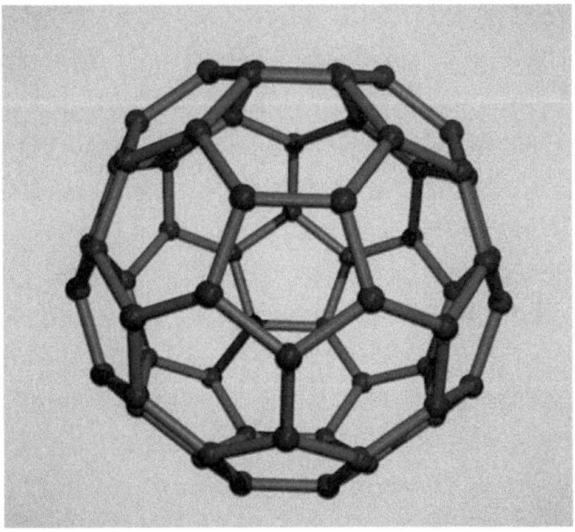

FIGURE 6.9 Schematic of fullerene.

A fullerene could be defined based on the number of hexagons possessed in addition to the fixed 12 pentagons (Fig. 6.9). There exist C_{70}, C_{59}, C_{84} fullerenes, to fullerenes with up to 300 carbon atoms, commonly called buckyballs or endohedral fullerenes.[62] The second type of fullerene is the one with more than 300 carbon atoms which is called the giant fullerenes which comprises single-shelled or multishelled carbon structures, carbon onions, etc.[62] The carbon nano-onions are multishelled fullerenes that are now being extensively used in FCs due to their properties of high surface area and high specific capacitance. It also possesses other properties of conductivity and tribology.[63]

Among all these, the most stable and widely used fullerene is the C_{60} fullerene due to its geodesic and electronic bonding factors. The attracting properties of this allotrope such as high electrochemical stability, small size, well-ordered structure, and specific morphology owes to its application in energy conversion systems.[63] Other properties of fullerenes include high stability and strong intermolecular interactions.[66] The most appreciable factor about the properties of fullerenes was that it could account for very unusual properties since it only had carbon atoms and fascinating cage-like structures.[66] Physical properties of C_{60} (fullerene) could be specified by mentioning its density: 1.65 g cm^{-3}, standard heat of formation: 9.08 k cal mol^{-1}, index of refraction: 2.2 (600 nm), boiling point: sublimes at 800 K, resistivity: 1014 ohms m^{-1}, crystal form hexagonal cubic, vapor pressure: 5×10^{-6} Torr at room temperature: 8×10^{-4} Torr at 800 K.[62]

Fullerenes are capable of forming composites with many polymers thus providing an added advantage to the fullerene derivatives by introducing the desired property. PCBM (Phenyl Cn Butyric Acid Methyl Ester) is an example of such a fullerene derivative. The other advanced forms of fullerenes include the dimer of fullerene that exists in dumb-bell- shape which is a connection of two fullerene cages.[64] Fullerenes could be modified into other forms such as thin films,[66] solid fullerene,[66] and could be used in solvents,[65] thus having a wide range structure.

6.4.1.1 *SYNTHESIS OF FULLERENE AND FULLERENE DERIVATIVES*

Research done till date shows that fullerenes could be synthesized by employing three different methods: (1) arc-discharge method, (2) laser-furnace method, and (3) sonophysical method of which the arc-discharge method is the most common and much-used method for producing bulk C_{60} fullerenes.

6.4.1.1.1 Arc-Discharge Method

The arc-discharge method for synthesizing fullerenes was first discovered and done by W. Kratschmer et al. in 1990 as a peak in the mass spectra of quenched carbon atoms. The carbon soot produced by evaporated graphite rods in contact was used as the precursor for producing bulk fullerenes.[66] Evaporation of the graphite rods was done using resistive heating in a glass jar that was filled with inert quenching gas.[70] Nucleation of the carbon vapor takes place thus forming smoke particles.[70] The produced smoke particles were collected on transparent substrates and gold-coated glass surfaces for transmission and reflection measurements, respectively.[70] The smoke collected on the transparent substrate would produce interference fringes on spectra taken at high resolution which could be avoided by using gold-coated glass surfaces. By conducting ultraviolet, infrared, and visible spectral studies at 100 Torr quenching gas pressure, different features in the above-mentioned absorption spectra were analyzed and the results confirmed the presence of strong bands representing pure carbon with an isotopic shift.[70] The results of the infrared spectra confirm that the features appeared as the weaker bands were due to the presence of large molecules or collection of large molecules which resembled C_{60} fullerenes. Studies also confirm the presence of C_{70} molecule which appear in a lesser amount. The confirmation of the presence of C_{60} fullerenes was done since the experimental result of the infrared spectra were in good agreement with the theoretical symmetry results.[70]

The C_{60} molecules produced were treated with ether for the removal of hydrogenated molecules, and the isolation of the C_{60} molecules was carried out by using toluene/hexane elutants.[66] The isolated molecules were further purified by employing sublimation at 623 K under 10^{-6} Torr thus obtaining 99.9% pure C_{60} molecules. The purity was confirmed by IR and UV mass spectroscopies.[66]

The synthesis method thus discovered by W. Kratschmer et al. was then modified by Yoshinori Ando et al. in 2004.[72] The latter group used a SiC powder production apparatus to provide a DC arc voltage between two separated graphite rods which served as two electrodes. The anode was then allowed to evaporate to produce fullerenes which were turned out as soot in the chamber.[72]

6.4.1.1.2 Laser-furnace method

Smalley and group was the pioneer to introduce the laser vaporization method which to synthesize fullerene.[80] The apparatus they used comprised of a pulsed laser beam focused on a rotating/translating graphite disk.[80] The intense laser beams were capable of producing superhot plasma that was then cooled in a helium atmosphere to generate clusters of carbon atoms. Thus, the process involves carbon condensation in a controlled and systematic way to produce large clusters of carbon atoms such as fullerenes.[72] Lasers always favor materials with high boiling temperatures such as carbon because the energy density of lasers is greater than any other vaporization device.[72] The difficulty encountered with this method was low quantity production that was overcome by introducing an annealing system together with the laser furnace method considering the fact that soccer ball structured fullerenes were synthesized at higher furnace temperature.[72]

6.4.1.1.3 Sonophysical Method Synthesis of Fullerenes

This method was used to synthesize big hollow nanobowls of C_{60} fullerenes.[63] This is a synthesis method that does not involve any chemical reactions. C_{60} fullerenes are mixed in acetonitrile:m-xylene in 1:3 ratio and big amounts of hollow nanobowls of were obtained when concentration was 1.4 mg mL^{-1}.[63] These nanobowls were used in a FC as an optically transparent electrode.

6.4.2 APPLICATIONS OF FULLERENE

The growing demand in the present world for a renewable energy substitute for the widely used fossil fuels due to the increasing pollution and high cost paved the way to find new alternatives in the field of energy conversion and storage. Thus, FCs, PV cells, and solar cells are some of the sustainable alternatives that could store and convert energy. Researches are widely being done to use fullerene or fullerene-based composite either fullerene derivatives in these energy conversion systems due to the high electrochemical stability, small size, well-ordered structure, and specific morphology.[63] The fullerenes could be used in solar cells and PV cells as an electron-accepting component due to their high electron affinity and superior ability to transport charge.[75]

6.4.2.1 FULLERENES IN FUEL CELLS (FCS)

FC is a device capable of converting chemical energy into electrical energy.[63] FC could work by feeding ethanol or methanol directly which has more efficiency over other heating systems.[63] Another type of FC is those working on a hydrogen base. These two existing types are encountering disadvantages that are the high cost for the former and difficulty in fuel production, storage, etc., for the latter.[63] And these difficulties are overcome by using liquid fuels specifically direct alcohol fuel cells.[63] Fullerenes are used as catalyst support in FCs due to their high specific surface area and porosity.[63]

Fullerenes or fullerene derivatives could be used in FCs in different aspects, such as fullerene derivatives enhancing the oxidation reaction at anode, used in reduction reaction at cathode, used in proton-conducting membranes,[63] and used as a catalyst for FCs.[74]

Fullerene derivatives enhancing the oxidation reaction at the anode: direct methanol fuel cells (DMFCs) are FCs that work on the basis of methanol oxidation. These FC are found to use fullerenes in the catalytic process.[63] The commercialization of DMFC depends on improving the methanol electro-oxidation reaction taking place at the anode. This is achieved by the use of Pt as an electrocatalyst metal, but the very high cost of this metal is a great challenge to the researchers.[63] The catalyst that could be used at the anode should possess a high surface area and thus employed fullerene as a catalyst for methanol oxidation. Studies were done on three different systems out of which the naked Pt/C_{60}/Au was found to be most active in the methanol oxidation experiments. Studies on hybrid nanoparticles of Pt/C_{60} and Pt–Ru/C_{60} show that the addition of fullerene could improve the morphology of nanoparticle and thus act as an efficient catalyst for methanol oxidation reactions.[63] The increased efficiency is due to the affinity for C_{60} transition metals. In addition, the current density curve of methanol oxidation was increased by the addition of fullerenes to Pt.[63] A composite of C_{60} with polyaniline (PAni) was also used as an electrocatalyst in methanol oxidation reactions.[63] Furthermore, a modified hybrid of PAni-emeraldine base-doped C_{60} whiskers was obtained as it showed potential applications with large surface area. Another composite that turned out to be a potential catalyst for FC applications was Pt-supported mesofullerene.[63] Studies on ITO electrodes coated with single-crystalline C_{60} microstructures synthesized by the liquid–liquid interfacial precipitation method confirm the activeness of the electrode toward methanol oxidation due to the increased electrochemical active area.[63]

Direct formic acid fuel cells, an advanced FC than DMFC, which could work at room temperature faced difficulty in commercialization due to the high cost of the anode. This was overcome by incorporating fullerene with cyclodextrin as support.[63]

Carbon nano-onions (CNOs), mentioned earlier, are being combined with Pt as an anode. CNO was prepared by arc discharge method of graphite rods immersed in water. It was also synthesized by CVD and thermal annealing of nanodiamond particles.[63]

6.4.2.2 FULLERENE COMPOSITES EMPLOYED AS CATHODE MATERIALS

The use of fullerene derivatives for oxygen reduction reaction (ORR) is less. Metallofullerene, fullerene with one of the carbon atoms replaced by Pt or any other metal, is commonly used to enhance the ORR at the cathode.[63] The ORR was increased by the introduction of graphene-C_{60} hybrid.[63] Very few studies have been conducted in this section to date.

6.4.2.3 FULLERENES USED IN PROTON-CONDUCTING MEMBRANES

It is the property of electron affinity that makes the fullerenes capable to use it as potential proton conductors. Fullerenes can increase the acidity of the acidic group attached to it due to the above-specified property. Nafion, a prefluorinated polymer was a commonly used proton conductor, this polymer was combined with fullerene to form a fullerene derivative which showed a tremendous increase in proton conductivities.[73] The fullerene derivatives synthesized to be used as proton conducting membranes are C_{60}–Nafion, $C_{60}(OH)_{12}$–Nafion, $HC_{60}(CN)_3$–Nafion, $HC_{60}(CN)_3$–$C_{60}(TEO)_n$–Nafion, $C_{60}H_{18}$, $C_{60}H_2$, $C_{60}(PO_3H_2)_n$, SPEEK–sfu (sulfonated oxidized fullerene).[63]

6.4.2.4 FULLERENE DERIVATIVES USED AS A CATALYST

The fullerene is the most commonly used catalyst in polymer electrolyte membrane FCs (PEM fuel cells), which convert hydrogen and oxygen gases into electricity and water.[74] The Pt electrodes used in these FC encountered the problem of stability problems, such as Pt coarsening and erosion of carbon black.[74] These stability problems were overcome by using metallofullerenes as an alternative catalyst because fullerenes are stable enough to reduce carbon corrosion. The platinum heterofullerne ($C_{59}Pt$) and platinum exofullerene (PtC_{60}) are the two metallofullerenes used as a catalyst.[74]

6.4.2.5 FULLERENES IN PHOTOVOLTAICS

PVs refer to the conversion of solar energy into electrical energy. This is the most suitable alternative source for fossil fuels as solar energy is a great renewable energy source. Solar cells could be divided into two types, namely, OSCs and inorganic solar cells, of which the former is the excitonic solar cells (formation of excitons) and the latter is the silicon-based solar cells.[75,76] Inorganic solar cells have better efficiency, but their use is restricted due to their high cost.[76] While the difficulty arising in OSCs is the poor power conversion efficiencies.[76] The efficiency of these OSCs could be improved by selecting the correct donor–acceptor pair and device architecture.[75] The exciton dissociation is taking place between an electron donor and an electron acceptor, which are materials of different electron affinities.[75] The role of fullerenes in an OSC is to act as an electron-accepting component due to its large electron affinity and superior ability to transport charge.[75] Studies have reported the increase in efficiency of the bulk heterojunction (BHJ) (bicontinous composite of donor and acceptor phases) by introducing polymer donors and fullerene acceptors.

The BHJ solar cells based on fullerene derivatives could be studied based on donor–acceptor interaction, morphology, etc. All the mechanisms involved in the BHJ solar cells are controlled by the donor–acceptor composites.[75] Thus, we are interested in studying the interaction between the polymeric donors and fullerene acceptor[75] poly(3-hexylthiophene) (P3HT)–fullerene derivative[6,6]–phenyl–C61–butyric acid methyl ester (PCBM), porphyrin-fullerene based dyads, etc., are some of the BHJ solar cells synthesized earlier.[14,15]

Studies on donor–acceptor interactions: The efficiency of the solar cells could be increased by working on the electronic properties of the polymer–fullerene interactions to absorb a greater amount of light, produce free electrons, and transport charges to respective electrodes.[14] The reasons to consider fullerenes as the ideal acceptor for solar cells are: (1) they possess energetically deep-lying LUMO with high electron affinity,[75] (2) they are capable of stabilizing negative charges, (3) they have very high photoinduced charge transfer, (4) possess high electron mobility, (5) it possesses a constant electronic structure (does not depend on the functionalization).

Studies based on morphology: Morphology plays an important role in BHJ since their performance depends on the physical interaction of donor and acceptor. Morphology depends on the crystallinity and miscibility of the materials used. It also depends on the solvent that we chose and its annealing.[75] The most important BHJ solar cells currently in research are MDMO-PPV/PCBM and P3HT/PCBM combinations, out of which the most reliable combination is P3HT/PCBM. Studies based on solvents suggest that the use of toluene as a solvent improves the efficiency of the BHJ solar cells, and this efficiency was further increased by the introduction of chlorobenzene.[75] The morphology was achieved to a greater extend by employing thermal annealing.

Other composites used: C_{70} derivative of PCBM has been found to increase the efficiency of its C_{60} derivative due to its higher absorption in the visible region.[75] Another composite which increased the efficiency of the acceptor is poly[{N-dodecyl-2,5-bis(2-thienyl)pyrrole}-alt-{2,1,3-benzothiadiazole}] (PTPTB) mixed with PCBM in 1:3 ratio.[75] The porphyrin–fullerene combination of donor–acceptor is also of great interest due to the low reorganization energies of both polymer and fullerene.[76]

Another class of solar cells is the organic–inorganic hybrid PSCs, which use fullerenes as electron transport materials, interfacial modification materials, or trap state passivators to improve efficiency.[77] Some examples of perovskite fullerene heterojunction devices are FTO/TiO_2/$MAPbI_3$:$PC_{61}BM$/spiro-OMeTAD/Au, ITO/PEDOT:PSS/$MAPbI_3$:$PC_{61}BM$/$PC_{61}BM$/Ca/Al, and ITO/PEDOT:PSS/$MAPbI_3$:1D $PC_{61}BM$/$PC_{61}BM$/Ca/Al.[77]

Thus, fullerenes form a super beneficial acceptor for PV cells that could act as a renewable energy substitute for fossil fuel.

6.4.2.6 *FULLERENES IN SUPERCAPACITOR*

SCs are the widely used energy storage device in the present world.[78] They could be used in the vehicle industry, memory devices, electronic devices, etc., the high power density, extraordinary cycling stability, low cost, low internal resistance, fast charging–discharging dynamics, and high rate capability[78] of SC makes it far better over the lithium-ion batteries. SC could be of two types: EDLCs and PCs. The energy storage of SCs is increased to a great extent by using a hybrid of EDLC and PC. The challenge encountered by SC these days is the delivery of energy density, which is found to be overcome by incorporating carbon nanoporous electrodes of high specific capacitance.[78] The advantages of carbon materials are low cost, better cycling stability, and a wide operating voltage window. Fullerenes are the commonly used nanoporous carbon material in SC in storage and conversion of energy, as it has a high surface area and high porosity.[78]

Polyaniline is the commonly used polymer for the electrode in SCs due to its low cost, facile synthesis, different oxidation states, and high doping level.[79] The efficiency of PAni electrodes could be enhanced by increasing the specific capacitance, which could be achieved by incorporating the fullerene derivative PCBM into PAni.[79] Another composite used in SC is P3HT: PCBM[79] as it showed excellent capacitance retention.[79]

As mentioned in the earlier section, the morphology of fullerenes is an important factor that determines the properties of the fullerene molecule. Out of many methods, the liquid–liquid interfacial precipitation (LLIP) method is more efficient in producing fullerenes with zero to higher dimensions.[78] This method is also employed to synthesize mesofullerene that has many applications in energy conversion systems. The fullerene molecules, thus, formed could be modified into fullerene with high surface area by annealing at high temperatures.[78] Further, the carbon derivatives such as CNTs and nanorods derived from fullerene molecules were found to exhibit higher surface areas than their commercially available counterparts.[78]

Shrestha and coworkers produced mesoporous carbon microtubes with graphitic pore walls from mesoporous crystalline fullerene C_{70} microtubes by heating the latter at a high temperature of 2000°C.[78] The microtubes, thus, produced could be used as electrodes in SCs as they have properties of high surface area and high specific capacitance.[78] Surfactant/fullerene hybrid materials were yet another composite synthesized by LLIP that showed long cycle stability, and it was able to sustain high current density.

Thus, studies conducted all through the years suggest that nanoporous carbon produced from self-assembled fullerene structures could be used as SC electrode material due to their enhanced surface area and high specific capacitance.

KEYWORDS

- **energy storage/generation**
- **carbon nanotube**
- **graphene**
- **fullerene**
- **supercapacitor**
- **fuel cells**

REFERENCES

1. Marco, N.; Liu, J.; Vernon, K.; Motta, N. *Beilstein J. Nanotechnol.* **2016,** *7*, 149–196.
2. Choi, H.-J.; Jung, S.-M.; Seo, J.-M.; Chang, D. W.; Dai, L.; Baek, J.-B. *Nano Energy* **2012,** *1*, 534–551.
3. Wu, M.; Liao, J.; et al. *Chem. Asian. J.* **2020,** *15* (7), 995–1013.
4. Book chapter, 2020, doi.org/10.5772/intechopen.94155
5. Brownson, D. A. C.; Kampouris, D. K.; Banks, C. E. *J. Power Sour.* **2011,** *196*, 4873–4885.
6. Pumera, M. *Energy Environ. Sci.* **2011,** *4*, 668–674.
7. Feng, J.; Zheng, D.; Gao, X.; Que, W.; Shi, W.; Liu, W.; Wu, F.; Cao, X. *Front. Energy Res.* **2020,** *8*.
8. Zhu, J.; Yang, D.; Yin, Z.; Yan, Q.; Zhang, H. *Small* **2014,** 10.1002/smll.201303202
9. Osawa, E. *Kagaku (Kyoto, Jpn.)* **1970,** *25*, 854–863.
10. Bochvar, D. A.; Galpern, E. G. *Dokl. Akad. Nauk SSSR* **1979,** *209*, 610–612.
11. Kroto, H. W.; Heath, J. R.; O'Brien, S. C.; Curl, R. F.; Smalley, R. E. *Nature* **1985,** *318*, 162–163.
12. Iijima, S. *Nature* **1991,** *354*, 56–58.
13. Geim, A. K.; Novoselov, K. S. *Nat. Mater.* **2007,** *6*, 183–191.
14. Swager, T. M. *ACS Macro Lett.* **2012,** 1, 3–5; (c) Sun, Z.; James, D. K.; Tour, J. M. *J. Phys. Chem. Lett.* **2011,** *2*, 2425–2432.
15. Mijatovic, D.; Eijkel, J. C.; Van den Berg, A. Technologies for Nanofluidic Systems: Top-Down vs. Bottom-Up—A Review. *Lab on a Chip* **2005,** *5* (5), 492–500.
16. Das, S.; Sudhagar, P.; Kang, Y. S.; Choi, W. Graphene Synthesis and Application for Solar Cells. *J. Mater. Res.* **2014,** *29* (3), 299–319.
17. Munoz, R.; Gómez-Aleixandre, C. Review of CVD Synthesis of Graphene. *Chem. Vapor Deposit.* **2013,** *19* (10-11-12), 297–322.
18. Novoselov, K. S.; Geim, A. K.; Morozov, S. V.; Jiang, D. E.; Zhang, Y.; Dubonos, S. V.; ... Firsov, A. A. *Science* **2004,** *306* (5696), 666–669.
19. Geim, A. K.; Novoselov, K. S.; pp. 11–19, 2010.
20. Rao, C. E. E.; Sood, A. E.; Subrahmanyam, K. E.; Govindaraj, A. *Angew. Chem. Int. Ed.* **2009,** *48* (42), 7752–7777.

21. Gao, W.; Singh, N.; Song, L.; Liu, Z.; Reddy, A. L. M.; Ci, L.; ... Ajayan, P. M. *Nat. Nanotechnol.* **2011,** *6* (8), 496–500.
22. Zhu, Y.; Murali, S.; Cai, W.; Li, X.; Suk, J. W.; Potts, J. R.; Ruoff, R. S. *Adv. Mater.* **2010,** *22* (35), 3906–3924.
23. Kong, X. K.; Chen, C. L.; Chen, Q. W. *Chem. Soc. Rev.* **2014,** *43* (8), 2841–2857.
24. Sun, Z.; Fang, S.; Hu, Y. H. 3D Graphene Mater. *Chem. Rev.* **2020,** *120* (18), 10336–10453
25. Li, G.; Huang, B.; Pan, Z.; Su, X.; Shao, Z.; An, L. *Energy Environ. Sci.* **2019,** *12* (7), 2030–2053.
26. Zhang, K.; Zhang, L. L.; Zhao, X. S.; Wu, J. *Chem. Mater.* **2010,** *22* (4), 1392–1401.
27. Ambrosi, A.; Chua, C. K.; Bonanni, A.; Pumera, M. *Chem. Rev.* **2014,** *114* (14), 7150–7188.
28. El-Kady, M. F.; Shao, Y.; Kaner, R. B. *Nat. Rev. Mater.* **2016,** *1* (7), 1–14.
29. Ji, L.; Lin, Z.; Alcoutlabi, M.; Zhang, X. *Energy Environ. Sci.* **2011,** *4* (8), 2682–2699.
30. Pan, D.; Wang, S.; Zhao, B.; Wu, M.; Zhang, H.; Wang, Y.; Jiao, Z. *Chem. Mater.* **2009,** *21* (14), 3136–3142.
31. Ji, L.; Meduri, P.; Agubra, V.; Xiao, X.; Alcoutlabi, M. *Adv. Energy Mater.* **2016,** *6* (16), 1502159.
32. Chen, T.; Xue, Y.; Roy, A. K.; Dai, L. *ACS Nano,* **2014,** *8* (1), 1039–1046.
33. Lu, Q.; Chen, J. G.; Xiao, J. Q. *Angew. Chem. Int. Ed.* **2013,** *52* (7), 1882–1889.
34. Li, Q.; Horn, M.; Wang, Y.; MacLeod, J.; Motta, N.; Liu, J. *Materials* **2019,** *12* (5), 703.
35. Goel, A.; Kumar, M. **2020,** *7* (7), 3586–3594.
36. Rogers, J. A.; Someya, T.; Huang, Y. **2010,** *327* (5973), 1603–1607.
37. Yang, X.; Cheng, C.; Wang, Y.; Qiu, L.; Li, D. *Science* **2013,** *341* (6145), 534–537.
38. Mahmoudi, T.; Wang, Y.; Hahn, Y. B. Graphene and its Derivatives for Solar Cells Application. *Nano Energy* **2018,** *47*, 51–65.
39. Xu, H. In IOP Conference Series: Earth and Environmental Science, 2020, October, Vol. 585, No. 1, p. 012027. IOP Publishing.
40. Sun, X.; Li, X.; Zeng, Y.; Meng, L. *Int. J. Energy Res.* **2021,** *45* (7), 10808–10820.
41. Su, H.; Hu, Y. H. *Energy Sci. Eng.* **2021,** *9* (7), 958–983.
42. Czerniak-Reczulska, M.; Niedzielska, A.; Jedrzejczak, A. *Adv. Mater. Sci.* **2015,** *15* (4), 67.
43. Gupta, N.; Gupta, S. M.; Sharma, S. K. *Carbon Lett.* **2019,** *29* (5), 419–447.
44. Shoukat, R.; Khan, M. I. *Microsyst. Technol.* **2021,** 1–10.
45. Ebbesen, T. W. **1994,** *24* (1), 235–264.
46. Sun, L.; Wang, X.; Wang, Y.; Zhang, Q. *Carbon* **2017,** *122*, 462–474.
47. Ando, Y.; Zhao, X.; Sugai, T.; Kumar, M. *Growing Carbon Nanotubes. Materials Today* **2004,** *7* (10), 22–29.
48. Sun, L.; Wang, X.; Wang, Y.; Zhang, Q. *Carbon* **2017,** *122*, 462–474.
49. Wu, Y.; Wang, J.; Jiang, K.; Fan, S. *Front. Phys.* **2014,** *9* (3), 351–369.
50. Kiamahalleh, M. V.; Zein, S. H. S.; Najafpour, G.; SATA, S. A.; Buniran, S. *Nano* **2012,** *7* (02), 1230002.
51. Liu, X. M.; Huang, Z. D.; Oh, S.; Ma, P. C.; Chan, P. C.; Vedam, G. K.; ... Kim, J. K. *J. Power Sour.* **2010,** *195* (13), 4290–4296.
52. Lota, G.; Fic, K.; Frackowiak, E. *Energy Environ. Sci.* **2011,** *4* (5), 1592–1605.

53. Endo, M.; Hayashi, T.; Ahm Kim, Y.; Terrones, M.; Dresselhaus, M. S. *Philos. Trans. Royal Soc. A* **2004,** *362* (1823), 2223–2238.
54. Akbari, E.; Buntat, Z. *Int. J. Energy Res.* **2017,** *41* (1), 92–102.
55. Liu, Z.; Lin, X.; Lee, J. Y.; Zhang, W.; Han, M.; Gan, L. M. *Langmuir* **2002,** *18* (10), 4054–4060.
56. Kim, J.; Lee, S. W.; Carlton, C.; Shao-Horn, Y. *J. Phys. Chem. Lett.* **2011,** *2* (11), 1332–1336.
57. Hu, X.; Hou, P.; Liu, C.; Cheng, H. *Nano Mater. Sci.* **2019,** *1* (3), 156–172.
58. Centi, G.; Perathoner, S. *ChemSusChem* **2011,** *4* (7), 913–925.
59. Maarouf, A. A.; Kasry, A.; Chandra, B.; Martyna, G. J. *Carbon* **2016,** *102*, 74–80.
60. Burakova, E. A.; Dyachkova, T. P.; Rukhov, A. V.; Tugolukov, E. N.; Galunin, E. V.; Tkachev, A. G.; Ali, I. *J. Mol. Liq.* **2018,** *253*, 340–346.
61. Hsia, B.; Marschewski, J.; Wang, S.; In, J. B.; Carraro, C.; Poulikakos, D.; ... Maboudian, R. *Nanotechnology* **2014,** *25* (5), 055401.
62. Yadav, B. C.; Kumar, R. *Int. J. Nanotechnol. Appl.* **2008,** *2* (1), 15–24.
63. Coro, J.; Suarez, M.; Silva, L. S.; Eguiluz, K. I.; Salazar-Banda, G. R. *Int. J. Hydrog. Energy* **2016,** *41* (40), 17944–17959.
64. Zhang, R.; Murata, M.; Wakamiya, A.; Murata, Y. *Chem. Lett.* **2013,** *42* (8), 879–881.
65. Ginzburg, B. M.; Tuichiev, S. *J. Macromol. Sci. B: Phys.* **2005,** *44* (4), 517–530.
66. Tanigaki, K.; Kuroshima, S.; Ebbesen, T. W. *Thin Solid Films* **1995,** *257* (2), 154–165.
67. Goel, A.; Howard, J. B.; Vander Sande, J. B. *Carbon* **2004,** *42* (10), 1907–1915.
68. Yang, M. Q.; Zhang, N.; Xu, Y. J. *ACS Appl. Mat. Interf.* **2013,** *5* (3), 1156–1164.
69. Parker, D. H.; Wurz, P.; Chatterjee, K.; Lykke, K. R.; Hunt, J. E.; Pellin, M. J.; ... Stock, L. M. *J. Am. Chem. Soc.* **1991,** *113* (20), 7499–7503.
70. Krätschmer, W.; Fostiropoulos, K.; Huffman, D. R. *Chem. Phys. Lett.* **1990,** *170* (2–3), 167–170.
71. Altman, E. I.; Colton, R. J. *Surf. Sci.* **1992,** *279* (1–2), 49–67.
72. Ando, Y.; Zhao, X.; Sugai, T.; Kumar, M. *Mater. Today* **2004,** *7* (10), 22–29.
73. Wang, H.; DeSousa, R.; Gasa, J.; Tasaki, K.; Stucky, G.; Jousselme, B.; Wudl, F. *J. Memb. Sci.* **2007,** *289* (1–2), 277–283.
74. Gabriel, M. A.; Deutsch, T.; Franco, A. A. *ECS Transac.* **2010,** *25* (22), 1.
75. Thompson, B. C.; Fréchet, J. M. *Angew. Chem. Int. Ed.* **2008,** *47* (1), 58–77.
76. Imahori, H.; Fukuzumi, S. *Adv. Funct. Mater.* **2004,** *14* (6), 525–536.
77. Deng, L. L.; Xie, S. Y.; Gao, F. *Adv. Electron. Mater.* **2018,** *4* (10), 1700435.
78. Shrestha, R. G.; Maji, S.; Shrestha, L. K.; Ariga, K. *Nanomaterials* **2020,** *10* (4), 639.
79. Ramadan, A.; Anas, M.; Ebrahim, S.; Soliman, M.; Abou-Aly, A. I. *Int. J. Hydrog. Energy* **2020,** *45* (32), 16254–16265.
80. Smalley, R. E. *Account. Chem. Res.* **1992,** *25* (3), 98–105.

CHAPTER 7

GREEN ENERGY APPLICATIONS OF GRAPHENE

VIDYA L.[1], APARNA RAJ[1], NEELIMA S.[1], RIJU K. THOMAS[2], and C. SUDARSANAKUMAR[1,*]

[1]*School of Pure & Applied Physics, Mahatma Gandhi University, Kottayam, Kerala, India*

[2]*Bharata Mata College, Thrikkakara, Ernakulam, Kerala, India*

[*]*Corresponding author. E-mail: c.sudarsan.mgu@gmail.com*

ABSTRACT

Human society is facing a worldwide energy crisis and environmental issues related to conventional energy production methods. To overcome these issues, several researches have been going on, all over the world, making the effective use of green energy sources for the production of clean and sustainable energy. Solar radiation, wind, water, green hydrogen, organic waste, etc., are green energy sources that are readily available and among them, solar energy is being mostly exploited for power generation. Energy production by the effective use of these green sources will help to fulfill the energy needs. Potential materials are sufficient for the power production needed to be identified carefully. Exploration of graphene-based materials in the last decade could identify them as potential candidates to use in different stages of energy production when compared with both organic and inorganic materials. Currently, to improve the efficiency of energy production, energy storage, conversion, and transport, graphene-based

Nanostructured Carbon for Energy Generation, Storage, and Conversion. V. I. Kodolov, DSc, Omari Mukbaniani, DSc, Ann Rose Abraham, PhD, and A. K. Haghi, PhD (Eds.)

materials are effectively used in every stage. This chapter is a modest attempt to present a general readership about green energy applications of graphene.

7.1 INTRODUCTION

Carbon, a nonmetallic tetravalent element, belongs to group 14, usually named the carbon family of the periodic table. When we look around, from the paper, we write to every cell of our body relies on the backbone of carbon. It has several forms, each having a unique structure and physical properties. The conventional allotropes, diamond and graphite, both have covalent bonds to share their valence electrons with neighboring carbon atoms resulting in a high melting point. Graphite, being the most common form, had piled sheets of carbon forming hexagonal structure[1] with a lower density than diamond due to the presence of large space between the layers.[2] Relatively new synthetic allotropes such as fullerene (buckminsterfullerene C_{60}) and carbon nanotubes have been discovered. Carbon exists in sp, sp^2, and sp^3 hybridizations and its allotropes are zero-dimensional sp^2 fullerenes, the two-dimensional sp^2 honeycomb lattice of graphene, or three-dimensional sp^3 diamond crystals, see Figure 7.1.

It came to be a surprise to the physics community when Andre Geim and Konstantin Novoselov and their collaborators in 2004 showed that a single sheet of carbon atoms can be isolated from graphite and electrical characterizations can be done on a single layer.[1] The Nobel Prize (Physics 2010) was awarded collectively to Andre Geim and Konstantin Novoselov "for groundbreaking experiments regarding the two-dimensional material graphene."[3] They worked in the production, isolation, identification, and characterization of graphene.[4]

The year 2018 witnessed that researchers from the Massachusetts Institute of Technology stalked two layers of graphene with one atom thickness and twisted one lightly, inducing insulating and superconducting properties that are absent in single graphene sheets.[5] These phases have similarities with high-temperature cuprite-based superconductors. It gave birth to a new field called twistronics that can evoke novel electronic devices.[5]

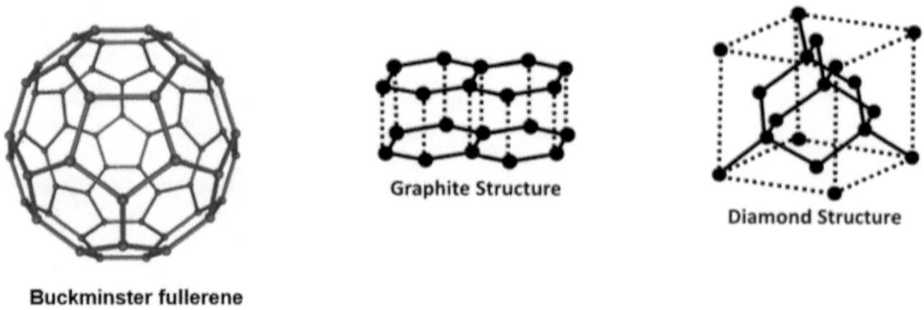

FIGURE 7.1 Allotropes of carbon. Reprinted with permission.

7.2 INTRODUCTION TO GRAPHENE

Graphene, theoretically first predicted by Wallace in 1947[6] is a 2D graphite, where carbon atoms are bonded through 6 sigma bonds to form a honeycomb structure.[7,8] Graphene is transparent, stronger than steel, and flexible as rubber that makes them the so-called super material.[9,10] The layers of carbon in graphite are stalked by weak Vander Waals forces. Each of these layers together forms the honeycomb structure, called graphene.[7] The high electron mobility of graphene is due to the appearance of delocalized electrons that are moving freely throughout the layers.[2] Graphene can be single-layered having 2D hexagonal sheets of carbon atoms or bilayered with two layers or with 3–10 layers of 2D sheets. The stalking in bilayered or few-layered graphene can be AA stalking, AB stalking, and rhombohedral or ABC stalking. Graphene has three sigma bonds/atoms *in-plane* and perpendicular pi-orbitals. The rigid backbone is formed by the sigma bonds while out-of-plane pi bonds engage in the interaction between different layers of graphene.[11] The stacking structure of graphene is well explained in Figure 7.2.

Graphene oxide (GO) is one of the successful graphene-based nanomaterials, due to its low production cost. It contains an oxygen functional group at high density. The oxidation of graphene lowers its electronic and mechanical properties, but forms stable suspensions in aqueous media. It also opens up the scope for a wide variety of chemical functionalization on GO sheets.

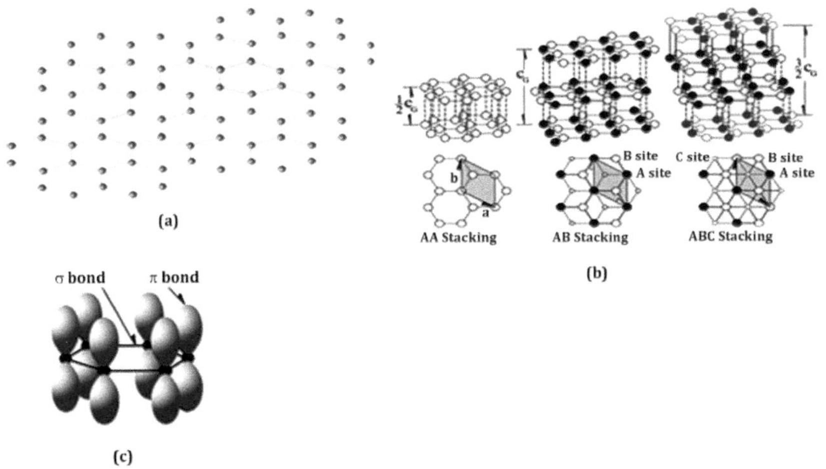

FIGURE 7.2 (a) Two-dimensional hexagonal graphene layers, (b) common structures and stacking structures of graphene, (c) schematic of the in-plane σ bonds and the π orbitals perpendicular to the graphene sheets plane.

Source: Reprinted with permission from Ref. [11]. © 2010 Taylor & Francis.

7.2.1 ELECTRONIC STRUCTURE AND PROPERTIES OF GRAPHENE

The electronic and electrical properties of graphene are the most underlying factors leading to the graphene revolution and they make them likely candidates for numerous applications. Their characteristics rely on the number of layers in the graphene sheets. Some of the features of graphene are given in Figure 7.3.[2]

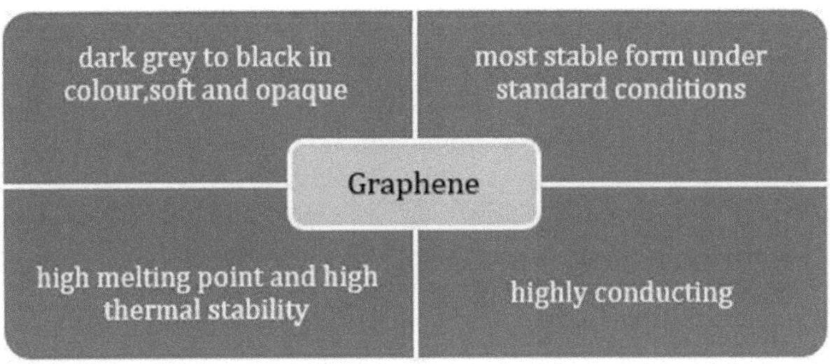

FIGURE 7.3 Basic features of graphene.

The flexibility of graphene is related to its electronic properties. The sp^2 hybridization between 1s and 2p orbitals guides to a trigonal planar structure, forming bonds between carbon atoms separated by 1.42 Å.[12]

Generally, carbon-based materials show excellent mechanical properties.[13] They can be used as follows:

- as a super strong structural material,
- to study and control the durability of graphene, which is beneficial in electronics and energy storage,
- to create curved graphene parts for electronics and structural applications, and
- to exploit nanocomposite with graphene inclusions as structural or functional materials.

The high optical transparency of graphene attracts the research interest. The linear dispersion of the Dirac electrons permits ultrawideband tunability and leads to potential photonics and photoelectronic applications. Excellent nonlinear optical properties combined with a fast response and broad wavelength range make graphene an ideal saturate absorber in mode-locked ultrafast laser systems.[13] The thermal conductivity of graphene due to phonon transport and electronic thermal transport in nondoped graphene is negligible because of its low carrier density. Mostly, pure graphene sheets are unreactive. Surface functionalization can make them reactive, and thickness has a major role to play in its reactivity.[9]

A polymer composite is a multiphase material in which fillers are reinforced into a polymer matrix yielding better mechanical properties which cannot be obtained from components when they are alone. Graphene can be used in composites as fillers. The properties of these composites depend on the ratio of filler to matrix and bonding between filler and matrix. Fabrication processes have a major role in the determination of resultant properties. Some of the fabrication methods of graphene–polymer nanocomposite are in-situ polymerization, melt intercalation technique, solution mixing, etc.[10]

7.2.2 SYNTHESIS METHODS OF GRAPHENE

The synthetic routes are mainly exfoliation and cleavage, thermal chemical vapor deposition technique (CVD); plasma enhanced chemical vapor deposition techniques. Other methods mainly include chemical methods, thermal deposition of SiC, thermal deposition on other substrates, and unzipping CNTs. Major synthesis routes of graphene are explained in Figure 7.4.[11] Some of these methods are evolved for specific applications like high electrical conductivity and some others require functionalized graphene.[13] Among the methods stated here, CVD is the most basic method for synthesis of high-quality pristine graphene (PG). PG, a next-generation graphene material, studied extensively because of its wider applications, mainly in electrochemical energy storage.[9]

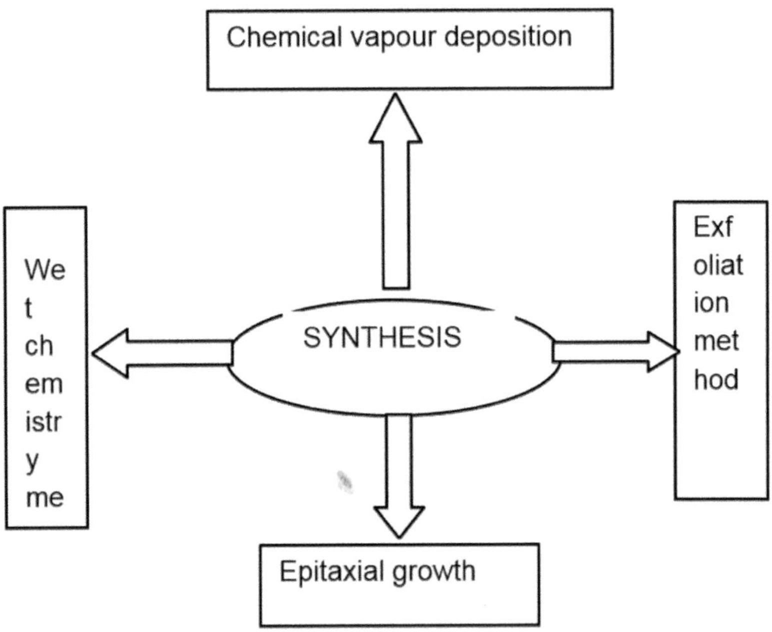

FIGURE 7.4 Major synthesis routes of graphene.

7.3 APPLICATIONS OF GRAPHENE

Detection of biomolecules is essential for disease diagnosis and therapy. Graphene biosensors are based on different sensing mechanisms like electrochemical signaling. Hydrogen peroxide (H_2O_2), being a moderator in biological processes its identification and detection is important. But its reactivity with electroactive substances interfere with the detection process. Here comes the importance of graphene-based electrodes that enhance the performance by increasing the electron transfer rate in comparison with graphite and base electrodes. Coming to the immune sensor, graphene is used because of its superior electrochemical properties and colossal surface area. GOs have fluorescence over a wide range of wavelengths, exhibiting interesting optical properties. This makes them useful in fabricating fluorescence resonance energy transfer sensors. GO shows excellent biocompatibility, flexible, and unique optical properties, which can be utilized in bioimaging applications. The large surface area makes graphene an efficient drug carrier to store considerable amounts of drug molecules on a single layer of the sheet. Besides chemotherapy and gene therapy, phototherapy (mainly photothermal therapy) and photodynamic therapy are used to treat diseases. Heat is generated by an optical absorbing agent under irradiation of light and the biological tissues, exposed to a high temperature favors selective elimination of abnormal cells.[15]

Environmental application mainly includes the detection of environmental pollutants such as heavy metals and toxic gases. Nowadays, graphene-based sensors are developed for their detection. PG shows extremely high optical transmittance and remarkable intrinsic charge mobility that has emerged as electrodes for chemical and biological sensing.[11] In photocatalysis, our aim is to produce highly efficient photocatalyst by combining photoresponsive material with graphene. Materials based on photocurrent generation offer potential applications in devices such as phototransistors, optical telecommunication, and photodetectors. A photodetector is a p–n junction that converts light to current. The low absorption cross-section and fast recombination rate of graphene limit its practical applications in photodetectors.[16]

7.3.1 GREEN ENERGY APPLICATIONS OF GRAPHENE

Renewable sources of energy are used for power generation, which do not give out greenhouse gases and thereby reduce pollution. That is the important reason behind the increased use of renewable energy sources based on fossil fuels. They mainly include solar, wind, biomass, etc. With respect to energy requirements, lots of efficient technologies are necessary for storage and generation. Nanotechnology has come up with a significant impact in this area. Among these nanotechnology materials, carbon-based materials have huge potential in energy applications.

7.3.1.1 SOLAR ENERGY

Over the past decades, the enormous development of technologies led to fast development in the field of energy. The energy produced in various forms plays an important role in the economic development and industrialization of the world. The continued use of fossil fuels has resulted in its depletion, pollution, and rise in cost. Also, environmental destruction is progressing due to the burning of fossil fuels, which results in the emission of greenhouse gases like CFCs, which is the major motive of ozone depletion and global warming. All these are the main reasons to find a solution that tries to reduce the use of fossil fuels to protect and boost the quality of life. Indeed, once traditional energy resources are exhausted, a serious energy crisis is foreseen. To meet global energy needs, we must seek out new technologies for the efficient fabrication, storage, and consumption of sources of energy. The generation of solar energy with semiconductor materials has been reported to generate clean energy. Solar is said to be a clean energy source because sunlight can be converted directly to generate electricity without pollution which would not have any direct hit on human life. Solar power is also one of the economical, secured, and purest renewable energy sources in the world.[14] To use solar power, a solar power generator and a photoelectric unit are needed. Using photovoltaic (PV) effects, PV materials can convert sunlight to electric power.

In contrast, solar power is on the throne of renewable energies such as wind, waves, and biomass. It could initiate a scientific revolution, and scientists will be encouraged to deepen their research on this subject. Some applications of solar energy will be noted in Figure 7.5. Recently, there were lots of studies that have been published on this topic .[18] Solar cells are very important devices for converting solar energy to electricity.[19] Using the technology of solar PV, solar radiant energy has been considered as one of the most abundant, limitless, and sustainable energy sources. Solar cells can convert up to 20% of incident solar radiation from solar radiation to electricity through the PV effect. Depending on the materials and fabrication process, solar cells are generally classified into three generations. Single crystalline silicon (Si) wafers and large polycrystalline

Si wafers are the former one of solar cells. These cells realize 12–16% solar conversion efficiency, depending on the manufacturing process and the quality of the wafer, and are assumed to be the leading material in the solar cell market.[7]

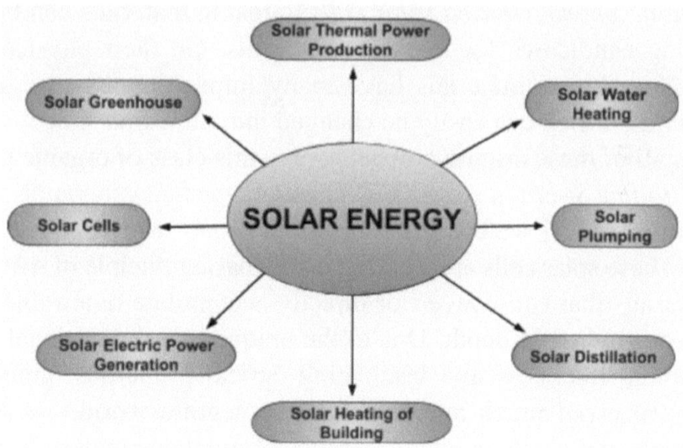

FIGURE 7.5 Applications of solar energy.

There are various kinds of solar cells available, among which metamorphic solar cells under standard solar spectrum ASTM G173 AM 1.5G possess the highest power-conversion efficiencies. Solar cells can be divided into different categories based on the type of materials and processing used in the manufacture of the devices.[18]

7.3.1.1.1 Types of Solar Cell

Thin film solar cell: These solar cells are made in thin films, for example, gallium arsenide, nanocrystalline Si, etc. Dye-sensitized solar cells (DSSCs) are one of another category belonging to thin film solar cells, which uses a different method to absorb sunlight from conventional methods. Solar dye-sensitized cells absorb sunlight through a semiconductor thin film.

Quantum dot (QD) solar cells: These are called multijunction solar cells based on the absorption of sunlight by QDs. QD solar cells have the potential to increase the maximum attainable thermodynamic conversion efficiency of solar photon conversion up to about 66% by utilizing hot photogenerated carriers to produce higher photovoltages or higher photocurrents.

QD solar cells have the capacity to raise the maximum possible thermodynamic conversion ability of solar photon conversion to around 66% by using hot generators to create higher voltages or fluorescence.

Organic polymer-based solar cells: Organic materials can be used as promising candidates for making solar cells. On their physical properties, organic-based materials have many improvements over inorganic compounds, which can easily be changed into ultra-thin sheets. Graphene satisfies all of these criteria and belongs to this class of organic matter.

Crystalline Si-based solar cells: These are the first generation solar cells from crystalline Si wafers, generally from mono- and poly-crystalline Si wafers. These solar cells are working on the basic principle of p–n junction.

Materials that can convert or directly accumulate renewable energies have been studied in depth. Due to the unique two-dimensional structure with atomic thickness and fascinating optical properties, graphene has recently received much attention in the scientific world.[21,22] Advanced mechanical and electrical characteristics of graphene materials as well as inexpensive and simple manufacturing processes enable them to be used as flexible conductive electrodes in capacitors, gas permeation barriers, or as light-emitting conductors in incandescent lamps.[23]

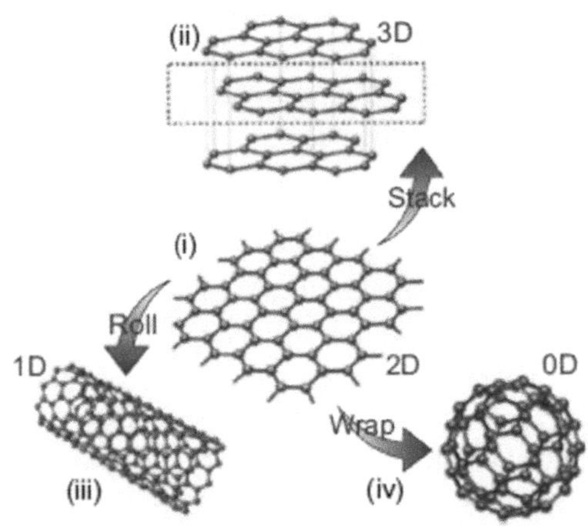

FIGURE 7.6 Hexagonal lattice present in carbon atoms of graphene (i), graphene sheets stacked into 3D graphite (ii), rolled into 1D nanotubes (iii), and wrapped into 0D Buckyballs. Reprinted with permissions.[24]

Depending on the properties of graphene, numerous graphene-based products have been developed.

As discussed before, graphene is an excellent material with many characteristics including structural, electronic, visible or optical, mechanical, and chemical properties. Unique atom thickness and high-level two-dimensional (2D) flatness result in a very high specific area and efficient electrical transport interfaces. The structural and electronic properties of GO offered a major role in developing the performance of solar cells. Due to the presence of highly delocalized electrons, the interaction with the P3OT is enhanced which results in a better donor–acceptor interface. In order to attain good and stable performance, chemically reduced graphene has been fused into organic solar cells. For example, GO convertible into an organic solution activated with penicillin isocyanate has been used as a new electron receptor material for poly (3-octylthiophene) (P3OT) organic solar cells.[25] Along with this, research in green energy has developed various ways, which includes chemical as well as physical methods. Also the hexagonal lattice atoms present the graphene will be represented in Fig. 7.6. Graphene and graphene-based products have been developed for the specific needs of humanity and for various applications in the research of green energy.[4]

7.3.1.1.2 Graphene-Based Polymer Solar Cells

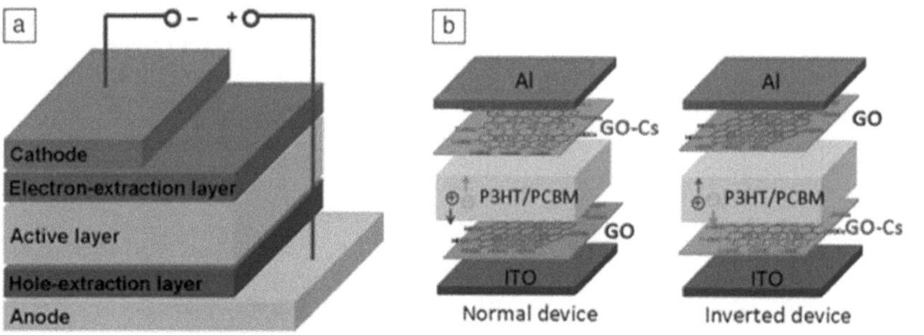

FIGURE 7.7 (a) Basic structure of polymer-based solar cell (PSC). (b) Normal (left) and inverted (right) PSCs with GO as the hole-extraction layer and cesium-neutralized graphene oxide (GO-Cs) as the electron-extraction layer.

Source: Reprinted with permission from Ref. [26]. © 2012 Springer Nature.

Polymer solar cell (PSC) is the one that offers advantages in its flexibility and cost and its basic structure is showed in Figure 7.7(a) and other type is in Figure 7.7(b). PSCs generally use an active layer made up of a mixture of donor, acceptor materials sandwiched between a cathode and an anode. After irradiation, electrons and holes are created by the transfer of photo-induced charges between the donor and the acceptor and are collected by the cathode and the anode. To accumulate these charges by electrodes, PSC requires an active layer for electron extraction. In recent days Indium tin oxide (ITO) is the widely used transparent electrode in PSC, but due to the limited availability of indium in the environment, ITO possesses some drawbacks like brittleness and high manufacturing costs. The graphene sheets manufactured by the chemical deposition method with excellent transparency offer a promising economic alternative for the former material (ITO electrode). Derivatives of graphene are also used as an active layer in PSCs. Layer-by-layer built graphene developed on a copper foil doped with an acid exhibits-high sheet resistance of 80 Q/square and excellent transmittance of 90% at 550 nm.[26]

7.3.1.1.3 Graphene–Inorganic Quantum Dot Hybrid Solar Cells

Inorganic QD solar cells are considered a better candidate for PV technology because they are able to overcome the limitations of the Shockley Queisser limit. But generally, QD solar cells have low PV efficiency due to poor electron-hole separation and poor photogenerated electron transfer to the electrode. These limitations can be solved by the use of QD solar cells with multilayered graphene as a novel electron acceptor. The properties like large specific area, adjustable band gaps, and good mobility of graphene help the resulting material to achieve 16% low incident photon-to-charge-carrier conversion efficiencies and show a photographic response of 10.8 A/m^2 under 1000 W/m^2 illumination.[26]

7.3.1.1.4 Graphene-Based Dye-sensitized Solar Cells

DSSCs have received considerable attention due to their moderate light-to-electricity conversion efficiency, especially due to its simple manufacturing process, and low cost. The counter electrodes perform an important role in reducing the redox species in DSSC by catalyzation.

It is used to transfer the flow of electrons from the outer circuit to the iodine and triiodide in the redox electrolytes. In DSSCs, one of the major requirements for acquiring high performance is the counter electrode to be highly catalytic and highly conductive. Graphene has been identified as an alternative solution to this problem. Unique graphene has also been found to have excellent electron transport properties. Until now, graphene has been theoretically studied extensively in relation to physics and electronic devices. Additionally, graphene has also been used to improve the quality of electrodes, supercapacitors, and field emission screens, along with this graphene can also be executed and developed for the applications in DSSCs.

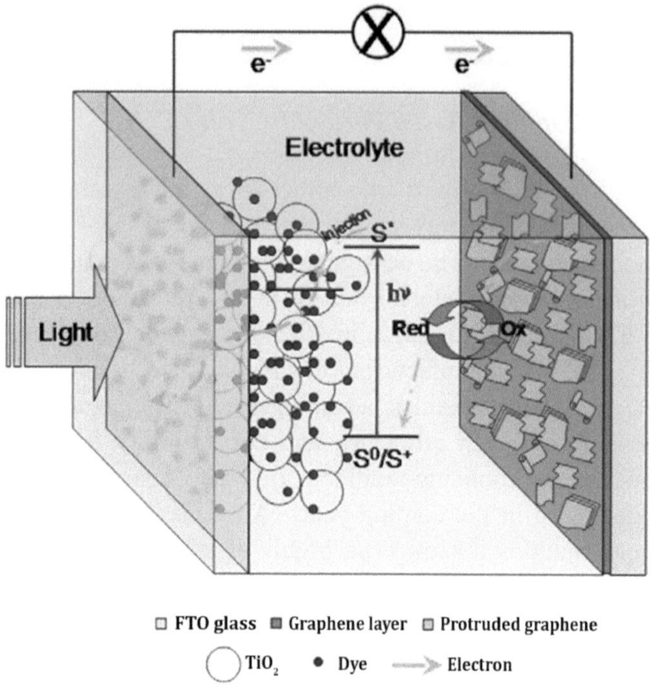

FIGURE 7.8 Schematic diagram of GM counter-electrode-based DSSC.
Source: Reprinted with permission from Ref. [27]. © 2012 Elsevier.

Graphene nanosheets are good counter electrode material for DSSC. However, the synthesis of the graphene involves some complex processing steps. Solar cell manufacturing is also a multistep process. Therefore, removing as many sources of pollution as possible during these multistep processes is very important. Impurities or defects act as recombination centers and can increase the dark current to reduce the fill factor of the device. In addition, graphene nanofilms have been found to aggregate easily due to their nanometric thickness. Aggregation reduces the electrochemical catalytic activity, which also causes an increase in the dark current. It has also been discovered that single graphene has fascinating electron transport properties. Until now, graphene has been widely studied theoretically in the phrase of physics and electronic devices. An increase in the purity and dispersion ability of graphene materials could increase the fill factor–thereby improving the efficiency of graphene-based DSSCs.[28] Schematic Diagram of GM counter-electrode-based DSSC is showed in Figure 7.8.

7.3.1.1.5 Graphene-Based Perovskite Solar Cells

Perovskite solar cells (PSCs) are another attracting research field because of their comparable energy conversion efficiency (ECE) to traditional commercial solar cells. The next-generation solar PV materials are PSCs. There is a widespread interest due to their rapid improvement in ECE reported 3.8% in 2009 to 22.1% in 2016. Other than that, its lightweight, flexibility, and low manufacturing costs also raise its importance in the scientific community. Besides, advantages of PSCs include, it can easily be fabricated with high efficiency and compatibility. Thus, the PSC is considered to be a promising candidate for energy conversion for coal-fired power generation in the coming years. Advances in PSC research offer unique opportunities for low cost, highly efficient, and thin film tandem solar cells. By combining the excellent PV properties, adjustable wide band gap and cold deposition technology of perovskite materials make them particularly attractive for multijunction tandem solar cells. Tandem solar cells can be manufactured with the advent of PSCs consisting of upper wideband perovskite cells and low band gap cells containing different materials, such as mixed Sn–Pb perovskite or crystalline silicon (c-Si), etc. Due to the standard power conversion efficiency of 23.7% and its low-cost manufacturing process, inorganic metal halide perovskite single-cross solar cells have gained a lot of priority in recent years. In addition to single-link devices, the adaptability of low-temperature solutions and band gap regulation make metal halide perovskite an ideal candidate for tandem solar cell production.[20] But the introduction of graphene-based materials helps to improve the costs, efficiency, chemical stability, energy level, etc. Moreover, the hydrophobic behavior of the surface of graphene provides protection against humidity in the ambient air, helping to improve the device life. By exploring the scope of graphene, new strategies can be developed to improve the performance and stability of devices before they can be industrialized.

7.3.1.2 HYDROGEN ENERGY

Think about an energy source that is a clean and perfect alternative to fossil fuels. It can be used irrespective of weather conditions unlike solar, wind, and hydropower. It should also have the potential to become the permanent solution to the world energy crisis and global environmental issues caused by fossil fuels. Of course, this is not an anecdote but the description of green hydrogen.[29–31]

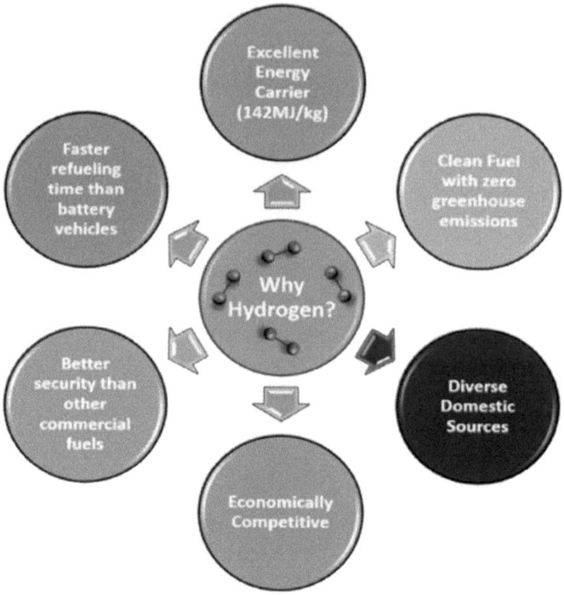

FIGURE 7.9 Advantages of hydrogen as a fuel.

Source: Adapted with permission from Ref. [31]. © 2019 Springer Nature.

Hydrogen energy is simply the energy produced using hydrogen and hydrogen-containing-compounds. Currently, more than 90% of the hydrogen is fabricated from natural gas or coal known as grey hydrogen and black hydrogen.[29] They are not green fuels since they contain high concentrations of carbon. These conventional methods are resulting in an inevitable carbon output accompanying the hydrogen. Green fuel-based hydrogen can be derived from green sources, such as water and organic waste. Green hydrogen can be used in a fuel cell to generate electric power with high efficacy without pollution of any kind. Production of hydrogen can be done by the electrolysis of water and is considered to be an environmentally friendly method when renewable energy sources are used and the whole process is cyclic. In addition to electrolysis, the production of hydrogen can be done using other processes like electro–thermo–chemical, electrochemical, photobiochemical, electrophotochemical. If the electrical energy required for these processes is generated from the energy contained in organic waste and renewable sources then the hydrogen produced will be green.[32]

For the development and effective use of hydrogen-based technologies, the storage and transport of hydrogen is significant. But these are the most challenging processes and have become a major problem for the commercialization of this fuel; therefore, exploitation of the hydrogen fuel for profit has not yet been fulfilled.[33,34] Compressing and liquefying the hydrogen is the conventional method for storage. These traditional methods endure from extreme processing conditions and include sizable safety risks. Later on, some carbon structures and metal hybrids are introduced to overcome the limitations of the earlier ones. But they were also not upto the mark for hydrogen storage. This is because no materials were able to perform high in terms of the storage capacity of hydrogen and the rate of adsorption/desorption processes. After that various studies about graphene have thrown a strong light on the viability of graphene structures for hydrogen storage devices.[31,35] The thermal and chemical stability enables the use of different synthesis methods for the production of scalable and economical graphene. There arises a new hope for the evolution of graphene-based solid-state porous materials, which shows promising applications in effective hydrogen storage. It was first realized in 2005 by using graphene for hydrogen storage.[36] Storage efficiency is usually governed by volumetric density and gravimetric density (wt % of hydrogen in total weight of graphene + hydrogen system).[35] Based on these parameter values, graphene becomes a perfect candidate for practical applications. Compared with other porous materials, graphene has a greater surface-to-mass ratio [due to the extremely high surface area (2630 m^2/g)]. This then reduces the problems related to small gravimetric densities. For designing efficacious materials, the interaction strategy of graphene and hydrogen should be known. Graphene absorbs hydrogen either by physisorption (van der Waals forces) or chemisorption (dissociative adsorption). Physisorption on graphene sheets is the first practically achieved method so it is studied more for a long time.[36] But due to weak binding energies and London dispersion forces, physisorption is more hard to control. Quantum fluctuation effects also non-negligible. On the other hand, the later introduced chemisorption method leads to more stably bonded hydrogen. Figure 7.10 represents the energy level diagram for the hydrogen–graphene system. The diagram shown here in accordance with energy in eV per H atom. So for finding the energy per hydrogen molecule (H_2), energy values in the diagram should be doubled. Tossini et al. calculated the value of barriers by theoretical evaluations. They took the unbound H_2+ PG as the reference level.[35]

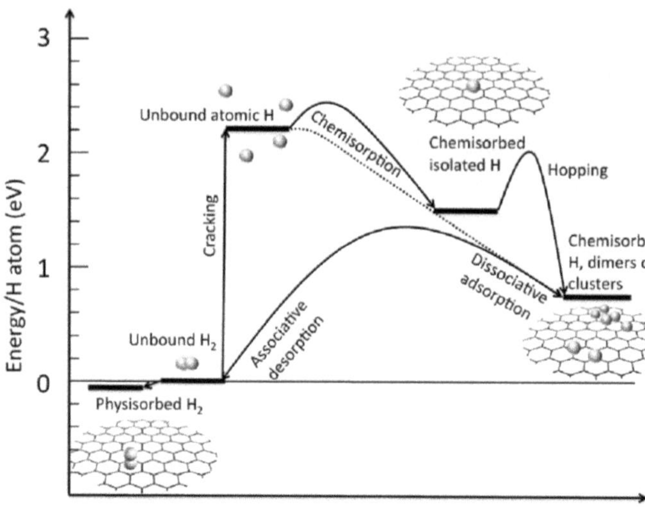

FIGURE 7.10 Energy level diagram for the interaction of graphene-based structure with hydrogen.

Source: Reprinted with permission from Ref. [35]. © 2013 Royal Society of Chemistry.

The favorable conditions for physisorption of the molecular hydrogen are low temperature and high pressure since the binding of H_2 molecules is very weak (theoretical value of binding energy is 0.01–0.06 eV).[34–36] In 2019, Bassem Assfour et al. reported that at 100 bars pressure and 77 K temperature highest hydrogen adsorption capacities were obtained for single-walled carbon tubes.[38] The appropriate tuning of some parameters like graphene curvature and axis of approach of hydrogen molecule will help to improve the adsorption binding energy which increases the storage stability at ambient temperature. The different sites of graphene in which physisorption occurs are between C–C bonds, on a carbon atom, or the hollow space in the middle portion of hexagon. According to the molecular axis of the approached hydrogen molecule, it can be adsorbed to graphene in any of the possible six configurations. The molecular axis can be parallel or perpendicular to the graphene sheet and can be adsorbed to any of the previously mentioned sites. This makes the possibility of six configurations. Theoretical studies/simulations revealed that the graphene can absorb hydrogen strongly through a parallel molecular axis approach compared to hydrogen with perpendicular axis approach. This may be due to the ellipsoidal electronic structure of hydrogen molecules.[36] Even though physisorption of hydrogen by graphene is a promising method, it suffers some major problems which restrict the practical applications. Physisorption at room temperature will be a bad idea since the molecule will be desorbed easily. Chemical functionalization and formation of chemical bonds between carbon and hydrogen are needed in order to achieve storage efficiency at room temperature. The p orbital present in the carbon atom in the graphene layer is the reason behind the amazing electrical properties and chemical reactivity. Chemisorption is the adsorption of hydrogen by forming a C–H chemical bond and changing the hybridization of carbon atom from stable sp^2 to sp^3.

Even though there is theoretical testimony for efficacious hydrogen storage by using graphene-based materials (doped, functionalized, or defected), there are many challenges for practical implementation. The major challenges are to identify the perfect combination of dopants and capping agent for graphene structures and optimal temperature for accurate results. Also, there are factors like modification of the surface (defects, curvature), usage of external fields, and optical stimuli, which will be needed to control and improve the storage kinetics of graphene.[31,36] Even then, we can definitely conclude that the new emerging technologies will ensure the development of appropriate porous graphene frameworks for commercialization of extraordinary hydrogen storage applications.

7.3.1.3 MICROBIAL FUEL CELL (MFC)

Microbial fuel cells (MFCs) are promising alternatives for the current energy harvesting systems with microorganisms that play the role as catalyst material. The cell converts the chemical energy of organic waste to electrical energy. Since MFCs are used for wastewater treatment and organic waste treatment effectively, it is an environmentally friendly technique.[38] A typical form of MFCs contains a cathode, an anode, and a membrane for H^+ exchange (proton exchange membrane). Figure 7.11 represents a schematic diagram of a two-chamber MFC.[39] The microorganism will produce a biofilm on the anode. When the wastewater is permitted to the anode chamber, the bacteria acts as it starts to oxidize the substrates and simultaneously produce electrons and protons. The electrons move through the external circuit to cathode and protons pass through the separating membrane from anode to cathode. A suitable electron acceptor like O_2 present in cathode to accept electrons. There are certain qualities of electrodes that will be the determining factors of efficiency of the fuel cell. MFCs should have a biocompatible anode and cathode material of excellent electrical conductivity.[40]

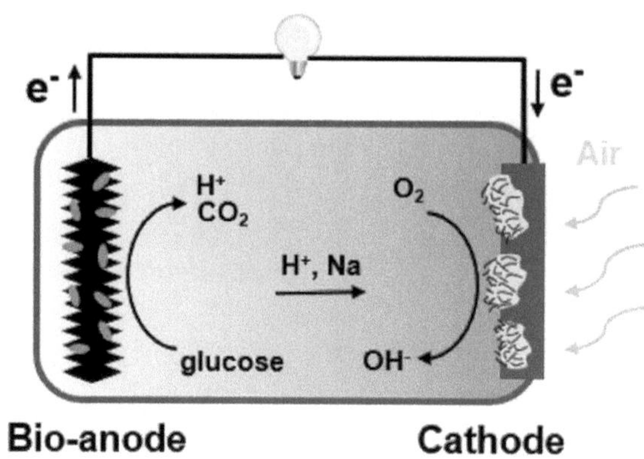

FIGURE 7.11 Schematic of the setup of microbial fuel cells.

Source: Reprinted with permission from Ref. [39]. © 2019 Elsevier.

The main drawbacks of practical application of MCFs technology are the low conversion efficiency, small charge transfer efficiency, low power density, and the high-cost cathode materials. To overcome these problems, many modifications are proposed in the MFC.[39] For this purpose, there were many materials like carbon-based materials and different metals used and studied. Recently, graphene-based materials are introduced into MCFs as catalysts and also in electrodes due to its amazing properties as large surface area, good conductivity, and feasible synthesis process. Platinum is the conventional cathode catalyst used in MCFs, but it is expensive. Recently, this conventional catalyst is replaced by inexpensive graphene. Also, several studies suggest that the charge transfer capacity of graphene is significantly high to be used as cathode material. The anode material should have high biocompatibility for bacterial colonization. Recently Alka Pareek et al. fabricated and characterized different three-dimensional graphene structure electrodes for studying the performance of MFCs.[41] They have synthesized efficient and cost-effective 3D graphene and characterized by X-ray and Raman spectroscopy. The study concludes that the 3D graphene is an efficient anode material for MFCs. Figure 7.12 represents the overall work of the team. In the fuel cell, graphene will form a graphene/biomolecule composite and in turn facilitate the charge transfer and improve the performance of this bioelectrochemical cell. Many of the problems regarding the commercialization of MFC can be solved by introducing graphene-based materials in the cell.

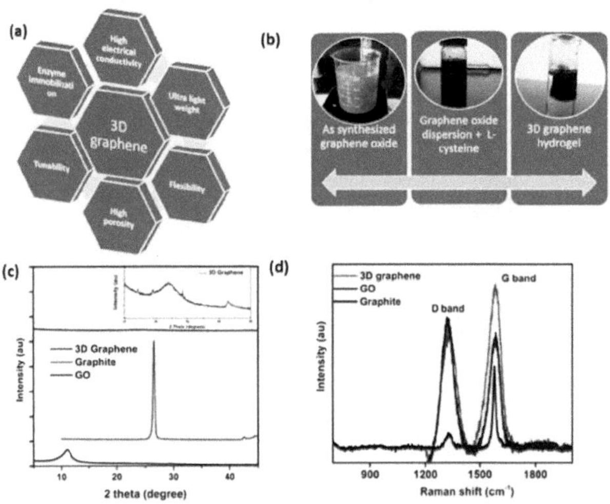

FIGURE 7.12 (a) Schematic diagram of advantages of 3D graphene, (b) images of different synthesized materials; (c) XRD and (d) Raman spectroscopy of graphite, GO, and 3D graphene.

Source: Reprinted from Ref. [41]. https://creativecommons.org/licenses/by-nc-nd/4.0/

7.3.1.4 WIND POWER AND HYDROPOWER

The process of generating electricity using the flow of air is referred to as wind energy and occurs naturally in earth's atmosphere. Indeed, the formation of wind is due to the heating effect of earth's surface unevenly as a result of the sun. Usually, modern technology uses wind turbines for the generation of kinetic energy. Mainly, there are three types of wind.

(1) Distribute wind (wind turbines < 100 KW and used for cottages and small fields without any connection to the electric grid).
(2) Utility-scale wind (size of the wind turbine is approximately about 100 KW to certain megawatts, here electricity is forwarded to the power grid and shared with the user by corresponding power operators).
(3) Offshore wind (possess large wind turbines).

Graphene can be transformed into a composite material to increase its strength without destroying the structure of the material. Graphene for nanomodification of wind turbine compounds provides wind power generation systems with promising wind turbine blades that are active, luminous, and more durable. In fact, graphene compounds like wind turbines are hardened epoxy compounds and outperform carbon nanotubes and other nanoparticle compounds. In general, epoxy composites demonstrate high tensile strength and good tensile modulus. Furthermore, graphene has the potential to reduce operating costs for wind turbines, mainly offshore wind. Batteries are also tried as an alternative power source, but these get quickly damaged and are not easy to replace. However, graphene can also serve as a coating material to protect advanced fiber-reinforced compounds from wind turbine blades from UV degradation.[42]

For a significant contribution to power supply, water can also be used as an alternative solution for solar and wind. Humanity has been drawing energy from water for thousands of years. Fortunately, or unfortunately, these hydrogenerators work efficiently only at large elevation differences and strong tides, so they do not utilize the energy of the water in intermittent droplets or streams. Thus, it is necessary to develop some advanced technologies for the easy conversion of low-quality hydroenergy from our surroundings into more useful electricity. There is a potential difference existing between the ends, when water flows via a medium that can be converted into electricity. It has been shown that moisture penetration, flow of solution, and evaporation of water and water–electrode interactions induce some unique property like hydropower generation because of water movement exists in various forms. Then, we made great strides in the operation and production of energy from graphene materials. Also, energy production from water–electrode interactions has recently been demonstrated and graphene-based composites have shown brilliant potential in fabricating self-powered devices guided by the interaction of water molecules with graphene. The electricity can be generated from the interaction of water and carbon materials and can also harvest power from graphene-based materials in accordance with ionic fluids and nonionic fluids.[43]

7.4 CONCLUSION

The ever-growing world population and the quality of day-to-day life require sufficient energy. Currently, human society depends on the sources of energy that will be exhausted in the nearby future and which is very harmful to the human environment. Hence, it is important to have secondary energy sources with less pollution, therefore green energy sources are identified as a perfect solution for all these problems. Nowadays, remarkable research has been carried out to implement energy-related materials to meet global energy demand. Materials that can directly convert or accumulate renewable energy have been extensively studied. Due to the unique and fascinating structural property of graphene, it can be used as a potential material to overcome energy demand in various ways. The chapter will provide an outlook to scientific reference for the application of graphene to green energy applications to achieve early large-scale energy conversion and storage.

ACKNOWLEDGMENT

Mrs. Neelima S. acknowledges the Department of Science and Technology, Government of India (DST-INSPIRE) for the award of JRF.

KEYWORDS

- **graphene**
- **green energy**
- **solar energy**
- **solar cell**
- **microbial fuel cell**
- **hydrogen energy**

REFERENCES

1. The Royal Swedish Academy of Sciences. Compiled by the Class for Physics of the Royal Swedish Academy of Sciences Graphene. 2010, 50005 (October), pp. 0–10.
2. Esquinazi, P. D., Ed. *Basic Physics of Functionalized Graphite: (Springer Series in Materials Science 244)*; Springer.
3. Grigorieva, I. V. Electric Field Effect in Atomically Thin Carbon Films. *Science* **2004,** *306* (5696), 666–9.
4. Castro Neto, A. H.; Guinea, F.; Peres, N. M. R.; Novoselov, K. S.; Geim, A. K. The Electronic Properties of Graphene. *Rev. Mod. Phys.* **2009,** *81*, 109.
5. Saito, Y.; Ge, J.; Watanabe, K.; Taniguchi, T.; Young, A. F. Independent Superconductors and Correlated Insulators in Twisted Bilayer Graphene. *Nat. Phys.* **2020,** *16*, 926–930.
6. Terrones, M.; Botello-méndez, A. R.; Campos-delgado, J.; López-urías, F.; Vega-cantú, Y. I.; Rodríguez-macías, F. J.; Laura, A.; Mu, E.; Cano-márquez, A. G.; Charlier, J.; et al. Graphene and Graphite Nanoribbons: Morphology, Properties, Synthesis, Defects and Applications. *Nano Today* **2010,** *5* (4), 351–372.
7. Spyrou, K.; Rudolf, P. An Introduction to Graphene. In *Functionalization of Graphene*; 2014.
8. Kharisov, B. I.; Kharissova, O. V. *Carbon Allotropes: Metal-Complex Chemistry, Properties and Applications*.
9. Du, W.; Geng, H.; Yang, Y.; Zhang, Y.; Rui, X.; Chao, C. Pristine Graphene for Advanced Electrochemical Energy Applications. *J. Power Sourc.* **2019,** *437* (June), 226899.
10. Phiri, J.; Gane, P.; Maloney, T. C. General Overview of Graphene: Production, Properties and Application in Polymer Composites. *Mater. Sci. Eng. B* **2017,** *215*, 9–28.
11. Choi, W.; Lahiri, I.; Seelaboyina, R.; Kang, Y. S.; Choi, W.; Lahiri, I.; Seelaboyina, R. Synthesis of Graphene and its Applications: A Review. *Crit. Rev. Solid State Mater. Sci.* **2010,** *35* (1), 52–71.
12. Korkmaz, S.; Af, İ. Graphene and Graphene Oxide Based Aerogels: Synthesis, Characteristics and Supercapacitor Applications. *J. Energy Stor.* **2020,** *27*, 101038.

13. Falkovsky, L. A. Optical Properties of Graphene Optical Properties of Graphene. *J. Phys.: Conf. Ser.* **2008,** 129, 012004.
14. Prabhakar, O.; Arun, K. Recent Progresses and Challenges in Graphene Based Nano Materials for Advanced Therapeutical Applications: A Comprehensive Review. *Mater. Today Communic.* **2020,** 22 (1), 100823.
15. Lakshmi, B.; Nouri, J. M.; Brabazon, D.; Naher, S. Graphene and Derivatives e Synthesis Techniques, Properties and their Energy Applications. *Energy* **2017,** *140,* 766–778.
16. Boehm, H.; Setton, R.; Stumpp, E. Nomenclature and Terminology Nomenclature and Terminology of Graphite. *Pure Appl. Chem.* **1994,** *66* (9), 1893–1901.
17. Singh, E.; Nalwa, H. S. Graphene-based Bulk-Heterojunction Solar Cells: A Review. *J. Nanosci. Nanotechnol.* **2015,** *15,* 6237–6278.
18. Herez, A.; Ramadan, M.; Abdulhay, B.; Khaled, M. Short Review on Solar Energy Systems. *AIP Conf. Proc.* **2016,** *1758,* 030041.
19. Lim, E. L.; Yap, C. C.; Jumali, M. H. H.; Teridi, M. A. M.; Teh, C. H. A Mini Review: Can Graphene Be a Novel Material for Perovskite Solar Cell Applications? *Nano-Micro Lett.* **2018,** *10,* 1–12.
20. Zhaoning Song, et al. Wide-Bandgap, Low-Bandgap, and Tandem Perovskite Solar Cells. *Semicond. Sci. Technol.* **2019,** *34,* 093001.
21. Tsang, C. H. A.; Huang, H.; Xuan, J.; Wang, H.; Leung, D. Y. C. Graphene Materials in Green Energy Applications: Recent Development and Future Perspective. *Renew. Sustain. Energy Rev.* **2020,** *120,* 109656.
22. Yusuf, M.; Kumar, M.; Khan, M. A.; Sillanpääillanpa, M.; Arafat, H. A Review on Exfoliation, Characterization, Environmental and Energy Applications of Graphene and Graphene-based Composites. *Adv. Colloid Interface Sci.* **2019,** *273,* 102036.
23. Scidà, A.; et al. Application of Graphene-based Flexible Antennas in Consumer Electronic Devices. *Mater. Today* **2018,** *21,* 223–230.
24. Lim, E. L.; Yap, C. C.; Jumali, M. M. H.; Teridi, M. A. M.; Teh, C. H. A Mini Review: Can Graphene be a Novel Material for Perovskite Solar Cell Applications? *Nano-Micro Lett.* **2018,** *10,* 27.
25. Guo, C. X., Guai, G. H., Li, C. M. Graphene Based Materials: Enhancing Solar Energy Harvesting. *Adv. Energy Mater.* **2011,** *1,* 448–452.
26. Liu, J.; Xue, Y.; Zhang, M.; Dai, L. Graphene-based Materials for Energy Applications. *MRS Bull.* **2012,***37* (12), 1265–1272.
27. Zhang, D. W.; Li, X. D.; Li, H. B.; Chen, S.; Sun, Z.; Yin, X. J.; Huang, S. M. Graphene-based Counter Electrode for Dye-Sensitized Solar Cells. *Carbon* **2011,** *49* (15), 5382–5388.
28. Zhang, D. W. et al. Graphene-based Counter Electrode for Dye-sensitized Solar Cells. *Car–bon N. Y.* **2011,** *49,* 5382–5388.
29. Low, F. W., Lai, C. W., Tiong, S. K., Amin, N. Graphene-Based Nanocomposites for Renewable Energy Application. In *Handbook of Polymer and Ceramic Nanotechnology*; 2019. doi:10.1007/978-3-030-10614-0_26-1.
30. Hoffman, J.; Hoffman, M. *Your Place in the New Energy Revolution.*
31. Jain, V.; Kandasubramanian, B. Functionalized Graphene Materials for Hydrogen Storage. *J. Mater. Sci.* **2020,** *55,* 1865–1903.

32. De Miranda, P. E. V.; Energy, E. H. Progress in Hydrogen Energy Infrastruc-Ture Development—Addressing Techni- Cal and Institutional Barriers. **2019**.
33. Assfour, B.; Leoni, S.; Seifert, G.; Baburin, I. A. Packings of Carbon Nanotubes—New Materials for Hydrogen Storage. *Avd. Mater.* **2011,** *23* (10), 1237–1241.
34. Bion, N.; Duprez, D.; Epron, F. Design of Nanocatalysts for Green Hydrogen Production from Bioethanol. *ChemSusChem* **2012,** *5*, 76–84.
35. Tozzini, V.; Pellegrini, V. Prospects for Hydrogen Storage in Graphene. *Phys. Chem. Chem. Phys.* **2013,** *15* (1), 80–89.
36. Patchkovskii, S.; Tse, J. S.; Yurchenko, S. N.; Zhechkov, L.; Heine, T.; Seifert, G. Graphene Nanostructures as Tunable Storage Media for Molecular Hydrogen. *PNAS* **2005,** *102* (30), 10439–10444.
37. Okamoto, Y.; Miyamoto, Y. Ab Initio Investigation of Physisorption of Molecular Hydrogen on Planar and Curved Graphenes. *J. Phys. Chem. B* **2001,** 105 (17), 3470–3474.
38. Liu, H.; Ramnarayanan, R. Production of Electricity during Wastewater Treatment Using a Single Chamber Microbial Fuel Cell. *Environ Sci Technol.* **2004,** *38* (7), 2281–2285..
39. Du, Y.; Ma, F.-X.; Xu, C.-Y.; Yu, J.; Li, D.; Feng, Y.; Zhen, L. Nitrogen-doped Carbon Nanotubes/Reduced Graphene Oxide Nanosheet Hybrids Towards Enhanced Cathodic Oxygen Reduction and Power Generation of Microbial Fuel Cells. *Nano Energy* **2019,** *61*, 533–539.
40. Joo, H.; Soo, H.; Sik, M.; Seop, I.; Kim, M.; Hong, B. A Mediator-Less Microbial Fuel Cell using a Metal Reducing Bacterium. *Shewanella Putrefaciens.* **2002,** *30*, 145–152.
41. Pareek, A.; Sravan, J. S.; Mohan, S. V. Fabrication of Three-Dimensional Graphene Anode for Augmenting Performance in Microbial Fuel Cells. *Carbon Resour. Convers.* **2019,** *2* (2), 134–140.
42. Low, F. W.; Lai, C. W.; Tiong, S. K. Amin, N. Graphene-based Nanocomposites for Renewable Energy Application. In *Handbook of Polymer and Ceramic Nanotechnology*; Hussain, C., Thomas, S., Eds.; Springer: Cham, 2019.
43. Han, Y.; Zhang, Z.; Qu, L. Power Generation from Graphene-Water Interactions. *FlatChem* **2019,** *14*, 100090.

INDEX

A

Adhesion, 163
Aerogel, 51
Aminated graphene honeycomb (AGH) structure, 53
Anodes, 194
Aqueous gel precursors, 8
Arc discharge
  method, 8, 201–202
  technique, 190
Arsenic (V) ions, 102–103
Assembly method
  energy storage applications
    characterization techniques, 19–21
    chemical vapor deposition (CVD), 17
    crystallinity and purity, 21
    electrical characterizations, 20
    layer-by-layer assembly method, 14–15
    radial breathing mode (RBM), 21
    solution processing, 11–13
    vacuum filtration, 15–16

B

Basal plane-functionalized, 68
Batch adsorption experiments, 100
Batteries
  heteroatom-doped/co-doped graphene applied, 134–135
  lithium–ion batteries (LIBs), 132
  lithium–oxygen (Li–O_2), 134
  nitrogen-doped graphene, 133
  rechargeable batteries, 132
  and supercapacitors, 193–194
  synergistic effect, 132
Bifunctional nickel ferrite decorated, 104–105
Bipolar plates, 182
Boron, 122
Boron doping, 127
Bottom-up method, 43
Brillouin zone, 148
Brunauer-Emmett-Teller (BET), 104
Bulk heterojunction (BHJ), 68

C

Carbon nanotubes (CNTs)
  catalysts by using
    adhesion in, 163
    diameter and number, 163
    dispersion of, 163
    electrode materials, use, 161
    with multimetal addition, 162
    with nitrogen, 161
    oriented CNT, 162
    surface treatment, 162
  chemical vapor deposition (CVD), 89
  electrochemical applications
    bifunctional nickel ferrite decorated, 104–105
    Brunauer-Emmett-Teller (BET), 104
    supercritical carbon dioxide ($SCCO_2$), 104
    transmission electron microscopy (TEM), 104
    Vertically Aligned Carbon Nanotubes (VACNTs), 104
    Zn–air batteries, 105
  electronic properties
    Brillouin zone, 148
    electron mobility, 148
    multiwalled CNTs (MWCNTs), 147
    single-walled carbon nanotubes (SWCNT), 147
  energy issues, 146–147
  ferrite nanocomposites, development
    hydrothermal method, 91
    methods, 90–91
    morphological analysis, 92
    nickel ferrite, 91
    TEM images, 92

ferrites, 89–90
fuel cells (FCs)
catalyst steadiness and corrosion resistance, 160
chemical energy, 159
decreases fuel cell cost, 160
high catalytic performance, 160
transmission capacity, 161
hydrogen storage, 155
adsorption and desorption, 157
applications, 157–159
chemisorption, 156–157
physisorption phenomenon, 156
lithium-ion batteries (Li-ion), 148
composite anode, 153–154
conventional anode materials, 151–153
energy requirements, 149
mechanism of, 150–151
merits and demerits as, 154–155
nanostructured materials, 149
rechargeable type, 150
magnetic data storage applications
cobalt ferrite ($CoFe_2O_4$), 105
X ray diffraction (XRD), 105
magnetic nanostructures, 88
metal oxide composite
Co_3O_4, 166
manganese oxide, 166
$Ni(OH)_2$, 167
ruthenium oxide, 165–166
microwave absorption
electromagnetic wave-absorbing properties, 93–95
energy-dispersive X-ray spectroscopy (EDS), 94
higher-order mode (HOM) load, 96
magnetic and reflection loss characteristics, 97–98
microwave sintering (MWS) technology, 97
multiwalled carbon nanotubes (MWCNTs), 95–96
nickel zinc ferrite (NZF), 95–96
radar absorption, 98–100
scanning electron microscopy (SEM), 94
sodium lignosulfonate (SLS), 97
strontium ferrite/CNT nanocomposites, 92–93
Vibrating Sample Magnetometer (VSM), 93
x-ray diffractometry (XRD), 97
photocatalysis
Co-doped copper ferrite, 106–107
Fourier transformation infrared (FTIR), 106
High Resolution Transmission Electron Microscopy (HRTEM), 107
multiwalled carbon nano tubes (MWCNTs), 106
photocatalytic performance, 107–108
solar-to-hydrogen (STH) conversion, 90
supercapacitors
heating effects, 165
structure effects, 164–165
synthetic conditions
low-temperature plasma technology, 164
microwave-assisted technology, 164
in supercritical fluid, 163
waste water treatment
arsenic (V) ions, 102–103
batch adsorption experiments, 100
cobalt ferrite nanoparticles, 100
Fourier transform infrared (FTIR) spectrometry, 101
Langmuir isotherm, 101
Raman spectroscopy, 101
rhodamine B (RhB), 100
Chemical vapor deposition (CVD), 89, 217–218
Chemically modified graphene (CMG), 47
Cobalt ferrite ($CoFe_2O_4$), 105
Cobalt ferrite nanoparticles, 100
Counter electrodes (CEs), 67
Cyclic voltammetry (CV), 49

D

Detection of biomolecules, 218
Doped electrodes, 57–58
Dry exfoliation, 177
Dye-sensitized solar cells (DSSCs), 28, 224–226

E

Edged carboxylated graphene nanoplatelets (ECGnPs), 68
Electric double-layer capacitor (EDLC), 45, 47, 129
Electrochemical applications
  bifunctional nickel ferrite decorated, 104–105
  Brunauer-Emmett-Teller (BET), 104
  supercritical carbon dioxide ($SCCO_2$), 104
  transmission electron microscopy (TEM), 104
  Vertically Aligned Carbon Nanotubes (VACNTs), 104
  Zn–air batteries, 105
Electronic properties
  Brillouin zone, 148
  electron mobility, 148
  multiwalled CNTs (MWCNTs), 147
  single-walled carbon nanotubes (SWCNT), 147
Energy storage applications
  assembly method
    characterization techniques, 19–21
    chemical vapor deposition (CVD), 17
    crystallinity and purity, 21
    electrical characterizations, 20
    layer-by-layer assembly method, 14–15
    radial breathing mode (RBM), 21
    solution processing, 11–13
    vacuum filtration, 15–16
  carbon nanotube (CNTs), 9
    one-dimensional (1D) nanotubes, 10
    with two-dimensional (2D), 10
  graphene–CNT hybrids, 5
    aqueous gel precursors, 8
    arc-discharge method, 8
    chemical vapor deposition method (CVD), 6
    consumable anodes, 8
    dye-sensitized solar cells, 28
    and field-effect transistors (FETs), 6
    flexible electrodes (TEs), 6
    in fuel cells, 21–23
    Hummer's method, 9
    perovskite solar cells, 27–28
    predispersed graphene oxide sheets, 9
    room temperature (RT), 7
    semi-transparent solar cells (SSCs), 29
    for solar cells, 26–27
    for supercapacitors, 23–25
    synthesis of, 6–9
    transparent, 5
  hybridization, strategies, 10
  nanomaterials, roles
    graphene, 4
    hole-transport layer (HTL), 4
    Indium tin oxide (ITO), 4
    multiwalled (MWCNTs) carbon nanotubes, 4
    oxygen reduction reaction (ORR), 4
    redox-based supercapacitors, 3
    single-walled (SWCNTs), 4
    supercapacitors, 3
    vertically aligned carbon nanotubes (VACNTs), 5
Energy-dispersive X-ray spectroscopy (EDS), 94

F

F-doped SnO2 (FTO), 70
Ferrite nanocomposites
  development
    hydrothermal method, 91
    methods, 90–91
    morphological analysis, 92
    nickel ferrite, 91
    TEM images, 92
Fourier transform infrared (FTIR) spectrometry, 101
Fuel cells (FCs)
  catalyst steadiness and corrosion resistance, 160
  chemical energy, 159
  decreases fuel cell cost, 160
  graphene exhibits, 74
  graphene-based carbon nitride (G-CN), 80
  heteroatoms, 79
  high catalytic performance, 160
  microbial fuel cell (MFC), 78

N-graphene with FePc, 77
nitrogen-doped (N-doped) graphene, 74
oxygen reduction reaction (ORR), 73, 75
PANI, 78
pliable paper, 77
polydiallyl dimethylammonium chloride (PDDA), 76
porous graphene with, 80
proton exchange membrane, 72
S-doped graphene, 76
transmission capacity, 161
VAGNAs and Pt NPs reinforced, 79
vertically aligned nitrogen-doped carbon nanotubes (VA-CNTs), 76
voltammograms, 76
Vulcan XC-72R carbon black catalyst, 73
XGnP electrodes, 80

G

Graphene
applications of
batteries, 182–183
as bipolar plates, 182
detection of biomolecules, 218
electrodes, 183
as electrodes, 182
environmental application, 219
fast charging batteries, 184
fuel cells (FCs), 181–182
hydrogen peroxide (H_2O_2), 218
li-ion batteries, 183
photodetector, 219
transparent electronics, 183–184
carbon nanotubes (CNT)
as anodes, 194
arc discharge technique, 190
batteries and supercapacitors, 193–194
chemical vapor deposition (CVD), 191
conductive additives, 195
current collectors, 195–196
fuel cells (FCs), 196–197
laser ablation technique, 191
lithium storage, 194
photovoltaics, 197–199
plasma-enhanced CVD (PECVD), 191
structure of, 191–193
synthesis and structural properties, 190
electronic structure and properties
features, 216
high optical transparency, 217
mechanical properties, 217
fullerene
applications of, 203
arc-discharge method, 201–202
cathode materials, 205
derivatives and synthesis, 201
fuel cells (FCs), 203–204
laser-furnace method, 202
in photovoltaics, 206–207
proton-conducting membranes, 205
sonophysical method, 202
in supercapacitor, 207–208
synthesis and structural properties, 199–201
used as a catalyst, 205
graphene oxide (GO)
structure, 179
green energy applications
crystalline si-based solar cells, 222
dye-sensitized solar cells (DSSCs), 224–226
hydrogen energy, 227–230
inorganic QD solar cells, 224
materials, 222
microbial fuel cells (MFCs), 231–232
nanotechnology, 219–220
organic polymer-based solar cells, 222
perovskite solar cells (PSCs), 226–227
polymer solar cell (PSC), 223–224
quantum dot (QD) solar cells, 221–222
renewable sources, 219
solar energy, 220–221
structural and electronic properties, 223
thin film solar cell, 221
wind power and hydropower, 233–234
supercapacitor (SCs)
components, 185
as electrode in, 186
flexible, 186
hole transport layer (HTL), 188–189
organic solar cells (OSCs), 188
perovskite solar cells (PSCs), 189–190
photovoltaics, 187
replacing batteries, 187

reversible faradaic redox reactions, 185–186
stretchable electronics, 186–187
synthesis and structural properties
CVD method, 178
3D graphene structures, 181
dry exfoliation, 177
facile method, 178
heteroatom-doped graphene, 181
Hummer's method, 178
liquid-phase exfoliation (LPE), 177
synthetic routes
chemical vapor deposition (CVD), 217–218
pristine graphene (PG), 218
Graphene nanoplatelets (GNP), 70
Graphene oxide (GO), 39, 215
batteries
heteroatom-doped/co-doped graphene applied, 134–135
lithium–ion batteries (LIBs), 132
lithium–oxygen (Li–O_2), 134
nitrogen-doped graphene, 133
rechargeable batteries, 132
synergistic effect, 132
environmental effect of, 81
fuel cells (FCs)
graphene exhibits, 74
graphene-based carbon nitride (G-CN), 80
heteroatoms, 79
microbial fuel cell (MFC), 78
N-graphene with FePc, 77
nitrogen-doped (N-doped) graphene, 74
oxygen reduction reaction (ORR), 73, 75
PANI, 78
pliable paper, 77
polydiallyl dimethylammonium chloride (PDDA), 76
porous graphene with, 80
proton exchange membrane, 72
S-doped graphene, 76
VAGNAs and Pt NPs reinforced, 79
vertically aligned nitrogen-doped carbon nanotubes (VA-CNTs), 76
voltammograms, 76
Vulcan XC-72R carbon black catalyst, 73
XGnP electrodes, 80
functionalization of, 39–40
heteroatom doping, 120
boron, 122
boron doping, 127
in energy storage applications, 128–129
graphitic type, 121
halogen atom doping, 128
hexagonal honeycomb lattice structure, 126
nitrogen bonding with, 121
nitrogen doping, 127
phosphorous doping, 127
pyrrolic nitrogen doping, 127
pyrrolic type, 121–122
reactive halogens, 122
sulfur doping, 128
synthesis methods, 122–126
literature survey
polycyclic aromatic hydrocarbons (PAHs), 41
single-layer graphene, 42
lithium-ion battery (LIB)
configuration, 56
doped electrodes, 57–58
graphene aerogel (GA), 60–61
graphene-based mesoporous silica (GM-silica) sheets, 60
graphene/SnO2-based nanoporous electrode, 59
graphitic carbon, 56
hydrogenated and halogenated, 63
$LiFePO_4$ (LFP), 60
novel tactic, 62
PF-grafted rGO (rGO-g-PF), 61, 63
rGO films, 60
sodium carboxy methyl cellulose (SCMC), 61
stacked platelet graphene nanofibers, 57
VAGNA electrodes, 61–62
self-assembly of, 40–41
solar cells
basal plane-functionalized, 68
bulk heterojunction (BHJ), 68
C_{60} grafted graphene (C_{60}-G), 65

counter electrodes (CEs), 67
edged carboxylated graphene nanoplatelets (ECGnPs), 68
F-doped SnO2 (FTO), 70
graphene film, 69
graphene nanoplatelets (GNP), 70
graphene-related materials (GRMs), 66–67
heteroatom-doped graphene, 71
hole transport layer (HTL), 70
incident photons to current efficiency (IPCE), 71
layer-by-layer stacked graphene, 64
N-doped graphene aerogel, 67
nitrogen edge-doped graphene nanoplatelets (NGnPs), 71–72
novel method, 67
organic photovoltaic (OPV) cell, 70
poly(3-hexylthiophene) (P3HT), 64
polyacrylic acid/cetyl trimethyl ammonium bromide (PAACTAB), 67
polymer solar cells (PSCs), 64
porphyrin-based photovoltaic structures, 71
power conversion efficiency (PCE), 64
processed poly(3-octylthiophene) (P3OT), 69
p-toluenesulphonyl hydrazine (p-TosNHNH$_2$), 70
PV devices, 66
semi-transparent perovskite solar cells, 72
structure and synthesis pathway, 65
transparent conducting oxide (TCO), 69
structural modification, 39
supercapacitors (SCs), 44
aerogel, 51
aminated graphene honeycomb (AGH) structure, 53
bandgap, 130
capacitance, 47–48
chemically modified graphene (CMG), 47
cyclic voltammetry (CV), 49
dielectric plates, 129
electric double-layer capacitor (EDLC), 45, 47, 129
Faradic reactions, 45
functioning of, 46
graphene/PANI multilayered nanostructure (GPMN), 52
layerby- layer (LBL) assembly techniques, 50
Nafion electrolyte membranes, 55
nitrogen-sulfur (N, S), 131
N-methyl-2-pyrrolidone (GPMN-N), 52
polyaniline (PANI), 49, 50
polyethylene imine-(PEI), 50
polymer/graphene nanocomposites, 55
specified surface area (SSA), 48
vertically aligned graphene nanosheet arrays (VAGNAs), 54
synthesis of
bottom-up method, 43
graphene sheets (GS), 43
large-scale production, 44
Graphene-based carbon nitride (G-CN), 80
Graphene-based mesoporous silica (GM-silica) sheets, 60
Graphene/PANI multilayered nanostructure (GPMN), 52
Graphene-related materials (GRMs), 66–67
Green energy applications
crystalline SI-based solar cells, 222
dye-sensitized solar cells (DSSCs), 224–226
hydrogen energy, 227–230
inorganic QD solar cells, 224
materials, 222
microbial fuel cells (MFCs), 231–232
nanotechnology, 219–220
organic polymer-based solar cells, 222
perovskite solar cells (PSCs), 226–227
polymer solar cell (PSC), 223–224
quantum dot (QD) solar cells, 221–222
renewable sources, 219
solar energy, 220–221
structural and electronic properties, 223
thin film solar cell, 221
wind power and hydropower, 233–234

H

Heteroatom doping, 120
- boron, 122
- boron doping, 127
- in energy storage applications, 128–129
- graphitic type, 121
- halogen atom doping, 128
- hexagonal honeycomb lattice structure, 126
- nitrogen bonding with, 121
- nitrogen doping, 127
- phosphorous doping, 127
- pyrrolic nitrogen doping, 127
- pyrrolic type, 121–122
- reactive halogens, 122
- sulfur doping, 128
- synthesis methods, 122–126

High Resolution Transmission Electron Microscopy (HRTEM), 107
Higher-order mode (HOM) load, 96
Hole transport layer (HTL), 4, 70, 188–189
Hummer's method, 9
Hydrogen storage, 155
- adsorption and desorption, 157
- applications, 157–159
- chemisorption, 156–157
- physisorption phenomenon, 156

I

Incident photons to current efficiency (IPCE), 71
Indium tin oxide (ITO), 4

L

Langmuir isotherm, 101
Layerby- layer (LBL) assembly techniques, 50
Liquid-phase exfoliation (LPE), 177
Lithium-ion batteries (LIBs), 132, 148
- composite anode, 153–154
- configuration, 56
- conventional anode materials, 151–153
- doped electrodes, 57–58
- energy requirements, 149
- graphene aerogel (GA), 60–61
- graphene-based mesoporous silica (GM-silica) sheets, 60
- graphene/SnO2-based nanoporous electrode, 59
- graphitic carbon, 56
- hydrogenated and halogenated, 63
- $LiFePO_4$ (LFP), 60
- mechanism of, 150–151
- merits and demerits as, 154–155
- nanostructured materials, 149
- novel tactic, 62
- PF-grafted rGO (rGO-g-PF), 61, 63
- rechargeable type, 150
- rGO films, 60
- sodium carboxy methyl cellulose (SCMC), 61
- stacked platelet graphene nanofibers, 57
- VAGNA electrodes, 61–62

Lithium–oxygen ($Li–O_2$), 134
Low-temperature plasma technology, 164

M

Metal oxide composite
- Co_3O_4, 166
- manganese oxide, 166
- $Ni(OH)_2$, 167
- ruthenium oxide, 165–166

Microbial fuel cells (MFCs), 78, 231–232
Microwave absorption
- electromagnetic wave-absorbing properties, 93–95
- energy-dispersive X-ray spectroscopy (EDS), 94
- higher-order mode (HOM) load, 96
- magnetic and reflection loss characteristics, 97–98
- microwave sintering (MWS) technology, 97
- multiwalled carbon nanotubes (MWCNTs), 95–96
- nickel zinc ferrite (NZF), 95–96
- radar absorption, 98–100
- scanning electron microscopy (SEM), 94
- sodium lignosulfonate (SLS), 97
- strontium ferrite/CNT nanocomposites, 92–93

Vibrating Sample Magnetometer (VSM), 93
x-ray diffractometry (XRD), 97
Microwave sintering (MWS) technology, 97
Microwave-assisted technology, 164
Multiwalled carbon nanotubes (MWCNTs), 95–96, 147

N

N-doped graphene aerogel, 67
N-graphene with FePc, 77
Nickel zinc ferrite (NZF), 95–96
Nitrogen edge-doped graphene nanoplatelets (NGnPs), 71–72
Nitrogen-doped (N-doped) graphene, 74
Novel method, 67

O

Organic photovoltaic (OPV) cell, 70
Organic solar cells (OSCs), 188
Oxygen reduction reaction (ORR), 4, 73, 75

P

Perovskite solar cells (PSCs), 27–28, 189–190
Photocatalysis
Co-doped copper ferrite, 106–107
Fourier transformation infrared (FTIR), 106
High Resolution Transmission Electron Microscopy (HRTEM), 107
multiwalled carbon nano tubes (MWCNTs), 106
photocatalytic performance, 107–108
Poly(3-hexylthiophene) (P3HT), 64
Polyacrylic acid/cetyl trimethyl ammonium bromide (PAACTAB), 67
Polyaniline (PANI), 49, 50
Polycyclic aromatic hydrocarbons (PAHs), 41
Polydiallyl dimethylammonium chloride (PDDA), 76
Polyethylene imine-(PEI), 50
Polymer solar cells (PSCs), 64
Power conversion efficiency (PCE), 64
Pristine graphene (PG), 218
Processed poly(3-octylthiophene) (P3OT), 69
P-toluenesulphonyl hydrazine (p-TosNHNH$_2$), 70

Q

Quantum dot (QD) solar cells, 221–222

R

Raman spectroscopy, 101
Redox-based supercapacitors, 3
Rhodamine B (RhB), 100
Room temperature (RT), 7

S

S-doped graphene, 76
Semi-transparent solar cells (SSCs), 29
Single-walled carbon nanotubes (SWCNT), 147
Sodium carboxy methyl cellulose (SCMC), 61
Sodium lignosulfonate (SLS), 97
Solar cells
basal plane-functionalized, 68
bulk heterojunction (BHJ), 68
C_{60} grafted graphene (C_{60}-G), 65
counter electrodes (CEs), 67
edged carboxylated graphene nanoplatelets (ECGnPs), 68
F-doped SnO2 (FTO), 70
graphene film, 69
graphene nanoplatelets (GNP), 70
graphene-related materials (GRMs), 66–67
heteroatom-doped graphene, 71
hole transport layer (HTL), 70
incident photons to current efficiency (IPCE), 71
layer-by-layer stacked graphene, 64
N-doped graphene aerogel, 67
nitrogen edge-doped graphene nanoplatelets (NGnPs), 71–72
novel method, 67
organic photovoltaic (OPV) cell, 70
poly(3-hexylthiophene) (P3HT), 64
polyacrylic acid/cetyl trimethyl ammonium bromide (PAACTAB), 67

polymer solar cells (PSCs), 64
porphyrin-based photovoltaic structures, 71
power conversion efficiency (PCE), 64
processed poly(3-octylthiophene) (P3OT), 69
p-toluenesulphonyl hydrazine (p-TosNHNH$_2$), 70
PV devices, 66
semi-transparent perovskite solar cells, 72
structure and synthesis pathway, 65
transparent conducting oxide (TCO), 69
Solar-to-hydrogen (STH) conversion, 90
Specified surface area (SSA), 48
Supercapacitors (SCs), 44
aerogel, 51
aminated graphene honeycomb (AGH) structure, 53
bandgap, 130
capacitance, 47–48
chemically modified graphene (CMG), 47
components, 185
cyclic voltammetry (CV), 49
dielectric plates, 129
electric double-layer capacitor (EDLC), 45, 47, 129
as electrode in, 186
Faradic reactions, 45
flexible, 186
functioning of, 46
graphene/PANI multilayered nanostructure (GPMN), 52
heating effects, 165
hole transport layer (HTL), 188–189
layerby- layer (LBL) assembly techniques, 50
Nafion electrolyte membranes, 55
nitrogen-sulfur (N, S), 131
N-methyl-2-pyrrolidone (GPMN-N), 52
organic solar cells (OSCs), 188
perovskite solar cells (PSCs), 189–190
photovoltaics, 187
polyaniline (PANI), 49, 50
polyethylene imine-(PEI), 50
polymer/graphene nanocomposites, 55
replacing batteries, 187
reversible faradaic redox reactions, 185–186
specified surface area (SSA), 48
stretchable electronics, 186–187
structure effects, 164–165
vertically aligned graphene nanosheet arrays (VAGNAs), 54
Supercritical carbon dioxide (SCCO$_2$), 104

T

Transmission electron microscopy (TEM), 104
Transparent conducting oxide (TCO), 69

V

Vacuum filtration, 15–16
Vertically aligned carbon nanotubes (VACNTs), 5, 104
Vertically aligned graphene nanosheet arrays (VAGNAs), 54
Vertically aligned nitrogen-doped carbon nanotubes (VA-CNTs), 76
Vibrating Sample Magnetometer (VSM), 93
Vulcan XC-72R carbon black catalyst, 73

W

Waste water treatment
arsenic (V) ions, 102–103
batch adsorption experiments, 100
cobalt ferrite nanoparticles, 100
Fourier transform infrared (FTIR) spectrometry, 101
Langmuir isotherm, 101
Raman spectroscopy, 101
rhodamine B (RhB), 100

X

X ray diffraction (XRD), 105
XGnP electrodes, 80
X-ray diffractometry (XRD), 97

Z

Zn–air batteries, 105